X-RAY DIFFRACTION FOR MATERIALS RESEARCH

From Fundamentals to Applications

X-RAY DIFFRACTION FOR MATERIALS RESEARCH
From Fundamentals to Applications

Myeongkyu Lee, PhD

Apple Academic Press Inc.	Apple Academic Press Inc.
3333 Mistwell Crescent	9 Spinnaker Way
Oakville, ON L6L 0A2	Waretown, NJ 08758
Canada	USA

©2016 by Apple Academic Press, Inc.

Exclusive worldwide distribution by CRC Press, a member of Taylor & Francis Group

No claim to original U.S. Government works

Printed in the United States of America on acid-free paper

International Standard Book Number-13: 978-1-77188-298-9 (Hardcover)

International Standard Book Number-13: 978-1-77188-299-6 (eBook)

All rights reserved. No part of this work may be reprinted or reproduced or utilized in any form or by any electric, mechanical or other means, now known or hereafter invented, including photocopying and recording, or in any information storage or retrieval system, without permission in writing from the publisher or its distributor, except in the case of brief excerpts or quotations for use in reviews or critical articles.

This book contains information obtained from authentic and highly regarded sources. Reprinted material is quoted with permission and sources are indicated. Copyright for individual articles remains with the authors as indicated. A wide variety of references are listed. Reasonable efforts have been made to publish reliable data and information, but the authors, editors, and the publisher cannot assume responsibility for the validity of all materials or the consequences of their use. The authors, editors, and the publisher have attempted to trace the copyright holders of all material reproduced in this publication and apologize to copyright holders if permission to publish in this form has not been obtained. If any copyright material has not been acknowledged, please write and let us know so we may rectify in any future reprint.

Trademark Notice: Registered trademark of products or corporate names are used only for explanation and identification without intent to infringe.

Library and Archives Canada Cataloguing in Publication

Lee, Myeongkyu, author
X-ray diffraction for materials research : from fundamentals to applications / Myeongkyu Lee, PhD.

Includes bibliographical references and index.
Issued in print and electronic formats.
ISBN 978-1-77188-298-9 (bound).--ISBN 978-1-77188-299-6 (pdf)
1. X-rays--Diffraction. 2. Materials--Analysis. 3. Crystal optics. 4. Crystallography. I. Title.

QC482.D5L43 2016	539.7'222	C2015-908302-8	C2015-908303-6

Library of Congress Cataloging-in-Publication Data

Names: Lee, Myeongkyu.
Title: X-ray diffraction for materials research : from fundamentals to applications / Myeongkyu Lee, PhD.
Description: Oakville, ON : Apple Academic Press, 2016. | Includes bibliographical references and index.
Identifiers: LCCN 2015047103 (print) | LCCN 2015048722 (ebook) | ISBN 9781771882989 (hardcover : alk. paper) | ISBN 9781771882996 () Subjects: LCSH: X-rays--Diffraction. | Materials--Research. Classification: LCC QC482.D5 L44 2016 (print) | LCC QC482.D5 (ebook) | DDC 620.1/1272--dc23
LC record available at http://lccn.loc.gov/2015047103

Apple Academic Press also publishes its books in a variety of electronic formats. Some content that appears in print may not be available in electronic format. For information about Apple Academic Press products, visit our website at **www.appleacademicpress.com** and the CRC Press website at **www.crcpress.com**

ABOUT THE AUTHOR

Myeongkyu Lee, PhD

Myeongkyu Lee, PhD, is a Professor in the Department of Materials Science and Engineering at Yonsei University in Seoul, Korea. He received the "best teacher award" from Yonsei University in 2006, and he has received best paper awards from the European Materials Research Society, the 10th IEEE Conference on Nanotechnology, and the American Conference on Crystal Growth. He teaches courses in crystallography, X-ray diffraction, optical properties of materials, advanced photonic materials, and opto-electronic properties of materials. He received his PhD in materials science and engineering from Stanford University, California, USA.

CONTENTS

List of Abbreviations .... *ix*

Preface .... *xi*

PART I: X-RAYS AND CRYSTAL GEOMETRY

1. **Electromagnetic Waves and X-Rays** 3

2. **Geometry of Crystals** 27

3. **Interference and Diffraction** 93

PART II: THEORY OF X-RAY DIFFRACTION

4. **Directions of X-Ray Diffraction** 117

5. **Intensities of X-Ray Diffraction** 147

PART III: APPLICATIONS OF X-RAY DIFFRACTION

6. **Characterization of Thin Films by X-Ray Diffraction** 181

7. **Laue Method and Determination of Single Crystal Orientation** 225

8. **Powder Diffraction** 249

Appendix. Fourteen Bravais Lattices 275

Bibliography 281

Index 283

LIST OF ABBREVIATIONS

BCC	body-centered cubic
Be	beryllium
FCC	face-centered cubic
FWHM	full width at half maximum
GIXRD	grazing incidence x-ray diffraction
HCP	hexagonal close-packed
ICCD	International Centre for Diffraction Data
IR	infrared
JCPDS	Joint Committee on Powder Diffraction Standards
KE	kinetic energy
PDF	powder diffraction file
TEM	transmission electron microscope
UV	ultraviolet

PREFACE

This book deals with the principle and applications of X-ray diffraction, which is a very useful and powerful tool for analyzing crystalline materials.

Since I joined Yonsei University as a faculty member in 2001, I have taught crystallography, X-ray diffraction, and optics to undergraduate students. X-ray diffraction is a consequence of the interaction between an electromagnetic wave and periodically arranged atoms. Thus, some knowledge on crystallography and optics is certainly helpful and, in a sense, requisite for a better understanding of the principle of X-ray diffraction. While working at the university, I have strongly felt the necessity of a textbook on X-ray diffraction for the beginners who do not have any background in crystallography and optics. My goal was to write a book that is easily accessible to undergraduates and consistently teachable. In writing this book, it was assumed that the potential readers have only an elementary knowledge of mathematics.

This book consists largely of three parts. As Part I, the first three chapters are given to explain the general properties of electromagnetic waves, the geometry of crystals, and the fundamentals of interference and diffraction. The rather lengthy Chapter 2 deals with basic crystallography that will be necessary to comprehend the underlying principle of X-ray diffraction and its applications. This chapter also covers the concept of lattice and reciprocal lattice, symmetry elements, crystal systems, and the crystal structures of some important materials, together with how the interplanar distances and angles in crystals can be determined. The theory of X-ray diffraction is described well in the Chapters 4 and 5 (Part II), where the direction and intensity of diffracted beams are discussed in detail. The final three chapters (Part III) describe how X-ray diffraction can be applied for characterizing such various forms of materials as single crystals, thin films, and powders. Thin film characterization is of scientific and technological significance, since modern electronic and optical devices are mostly based on thin films. Although a great number of research articles appear on thin film characterization by X-ray diffraction, there are few books solely focused on this topic. Therefore, a considerable portion of the application sections is devoted to thin film analysis. X-ray diffraction is a powerful nondestructive technique for characterizing thin films, providing a variety of information, such as phase, lattice parameter, film thickness, orientation, internal stress and strain, etc. The purpose of the relevant chapter is to introduce X-ray diffraction techniques that are widely used to characterize thin films deposited on substrates.

I hope that this book will be of use to the students in materials science, physics, and chemistry throughout their undergraduate and early graduate years.

Myeongkyu Lee
Department of Materials Science and Engineering
Yonsei University, 134 Sinchon-dong, Seoul, Korea
Office: B326, Engineering Building 2
Tel: +82-2-2123-2832
Fax: +82-2-312-5375
E-mail: myeong@yonsei.ac.kr

PART I
X-RAYS AND CRYSTAL GEOMETRY

CHAPTER 1

ELECTROMAGNETIC WAVES AND X-RAYS

CONTENTS

1.1	Materials Analysis by X-Ray Diffraction	4
1.2	Electromagnetic Spectrum	5
1.3	Wave-Particle Duality	12
1.4	Generation of X-Rays	15
1.5	Absorption	20
Problems		25

1.1 MATERIALS ANALYSIS BY X-RAY DIFFRACTION

X-rays refer to the electromagnetic radiations that have a wavelength range of 10^{-3} nm to 10 nm. X-rays, discovered by W. Röntgen in 1895, were so named because their characteristics were unknown at that time. Today, X-rays are widely used to image the inside of visually opaque objects, for example, in medical radiography, computed tomography, and security scanners. X-rays can also reveal various information on the materials, including crystal structure, phase transition, crystalline quality, orientation, and internal stress. This is made possible as a consequence of the interaction between X-rays and matter. X-rays with wavelengths below 0.1–0.2 nm are called *hard X-rays*, while those with longer wavelength are called *soft X-rays*. The X-rays utilized for materials analysis are hard X-rays. There are two reasons why X-rays are so powerful for analyzing the internal state of crystalline materials. First, hard X-rays are deeply penetrating into all substances, although the penetration depth varies with the substance. While metals are optically opaque, they may be transparent or translucent to the hard X-rays. Secondly, X-rays have much shorter wavelengths than visible light. This makes it possible to probe small structures that cannot be seen under an ordinary microscope. In particular, hard X-rays have wavelengths similar to the size of atoms. Therefore, they can be diffracted by atoms periodically arranged within the substance. Monitoring the diffraction direction and intensity allows the internal structure of crystalline matters to be revealed at the atomic level.

The diffraction of light had already been known before X-rays were discovered. Diffraction refers to various phenomena that occur when a wave encounters an obstacle or a slit. In classical physics, the diffraction phenomenon is described as the bending of waves around small obstacles and the spreading out of waves passing through small openings. Diffraction occurs with all waves, including sound wave, water wave, and electromagnetic waves, such as visible light, X-rays, and radio waves. It is well known that when a light wave is confronted with a periodic structure, it is split into several waves traveling in different directions. This behavior is also called diffraction, in which the periodic structure plays a role of diffraction grating. The diffraction effect becomes most profound when the wavelength is comparable to the grating period. Most materials are crystalline with regularly arranged atoms and X-rays have wavelengths similar to the interatomic distances. A crystalline matter contains many atomic

Electromagnetic Waves and X-rays

planes of different orientation and spacing, each of which can act as an effective diffraction grating when an X-ray beam is incident. The diffraction pattern, characterized by the direction and intensity of the diffracted beams, is characteristic of the matter and its internal structure. X-ray diffraction is a very powerful, nondestructive tool for analyzing materials and a variety of information can be deduced from the obtained diffraction pattern.

This book deals with the principle and applications of X-ray diffraction and the Chapters 1 and 2 are concerned with the general properties of electromagnetic waves and the geometry of crystals, respectively. X-ray diffraction is a consequence of the interplay between electromagnetic radiation and periodic atoms. Therefore, some knowledge on these topics is essential to comprehend the underlying principle of X-ray diffraction, which are described in Chapters 3, 4, and 5. X-ray diffraction is a versatile technique that can be utilized for phase identification, orientation determination, lattice parameter measurement, assessment of crystal quality, and determination of crystal structure. These common application areas are discussed throughout the final three chapters. Thin film characterization is of scientific and technological significance, since modern electronic and optical devices use many different thin films deposited on substrates. For optimal performance, these films are required to possess specific properties that are strongly affected by their microstructure. Therefore, characterization of thin film microstructure is very important to improve the device quality to an acceptable level. X-ray diffraction is an indispensable technique for nondestructive *in-situ* characterization of thin films. It can provide a variety of information such as phase, lattice parameter, film thickness, orientation relation with the substrate, internal stress and strain, etc. However, there are few textbooks solely concentrated on this topic. A considerable portion of the application sections is thus devoted to thin film characterization by X-ray diffraction.

1.2 ELECTROMAGNETIC SPECTRUM

Electromagnetic wave is a transverse wave propagating with oscillating electric and magnetic fields. In all waves, the displacement varies both with the position (x) and time (t). Thus, a wave traveling in the x-direction with a speed v satisfies the following one-dimensional differential wave equation

$$\frac{\partial^2 f}{\partial t^2} = v^2 \frac{\partial^2 f}{\partial x^2} \tag{1.1}$$

where $f(x,t)$ is the displacement at x and t. In an electromagnetic wave, $f(x,t)$ represents the electric or magnetic field. The electric and magnetic field components are orthogonal to each other and perpendicular to the direction of wave propagation. Since the magnetic field is always in phase with the electric field and its magnitude is proportion to the electric field strength, it is general to deal with the electric field only in optics and X-ray diffraction. Let's consider an initial electric field distribution shown in Figure 1.1(a). The arrows represent the direction and magnitude of the electric field. When the wave moves with a speed v, the electric field will have a different distribution like Figure 1.1(b) after a time t. If the spatial distribution at $t = 0$ is given by $f(x)$, it is represented by $f(x-vt)$ at $t = t$ because the wave propagates by a distance vt along the x-direction. Thus, $f(x\ t)$ is the most general form of the one-dimensional wave function. Of course, the wave function has the form of $f(x+vt)$ when the wave propagates in the negative x-direction. The electromagnetic wave is mathematically expressed by a sinusoidal function, as shown in Figure 1.1(c).

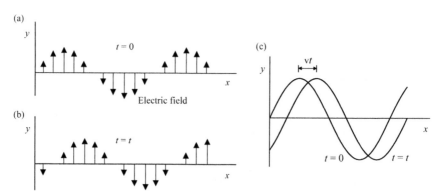

FIGURE 1.1 The variation of electric field with x at (a) $t = 0$ and (b) $t = t$. (c) The propagation of electromagnetic wave expressed by a sinusoidal function.

Although waves can have many different forms such as square, triangle, and saw-tooth, sinusoidal functions are the basic building blocks

Electromagnetic Waves and X-rays

representing waves. It follows from Fourier series that any periodic profiles can be generated by a superposition of sinusoidal functions. Electromagnetic wave is characterized by its *wavelength* (λ) or *frequency* (v) and this is equally applied to non-sinusoidal waveforms. The wavelength means the spatial period of the wave and the frequency is the inverse of the temporal period (Figure 1.2). The amplitude of electromagnetic wave corresponds to the maximum value of the electric field (E_o). Thus, when the electric field is directed along the *y*-axis, an electromagnetic wave propagating in the positive *x*-direction can be represented as follows

$$y = E_o \cos\left(\frac{2\pi}{\lambda}x - 2\pi v t\right) \text{ or } E_o \sin\left(\frac{2\pi}{\lambda}x - 2\pi v t\right) \quad (1.2)$$

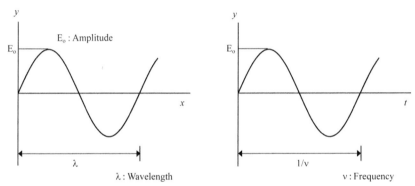

FIGURE 1.2 Definitions of the wavelength, frequency, and amplitude.

It does not matter whether the wave is expressed by a cosine function or a sine function because an arbitrary point can be taken as the origin in the moving wave. As will be discussed in Chapter 3, the behavior of waves can be better explained by describing it with a complex function.

The term in parentheses of Eq. (1.2) is known as the *phase* (θ) of the wave, so that

$$\theta = \frac{2\pi}{\lambda}x - 2\pi v t \quad (1.3)$$

The displacement of the wave is identical at the same phase. In Figure 1.3, two wave configurations have the same displacement at x_1, t_1 and x_2, t_2, which means that the phases at x_1, t_1 and x_2, t_2 are identical, i.e.,

$$\frac{2\pi}{\lambda}x_1 - 2\pi v t_1 = \frac{2\pi}{\lambda}x_2 - 2\pi v t_2 \tag{1.4}$$

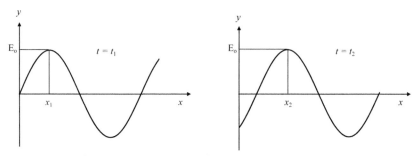

FIGURE 1.3 The phase and displacement of the wave at x_1, t_1 are the same as those at x_2, t_2.

It can be seen from Eq. (1.4) that the speed of the wave is equal to the product of its frequency and wavelength.

$$\frac{x_2 - x_1}{t_2 - t_1} = \frac{dx}{dt} = v = \nu \lambda \tag{1.5}$$

Since the electromagnetic wave moves at a constant speed of c = 2.998 × 10^8 m/s in vacuum, the wavelength is inversely proportional to the frequency, and vice versa. The electromagnetic radiation is classified by its wavelength or frequency into radio wave, microwave, infrared (IR) light, visible light, ultraviolet (UV) light, X-rays, and γ-rays. However, there are no sharp boundaries between the regions. Figure 1.4 shows the whole spectrum of the electromagnetic radiation. Our naked eyes can sense a relatively small range of wavelengths (400–700 nm) called visible spectrum or simply light. Other wavelengths, especially nearby IR (longer than 700 nm) and UV (shorter than 400 nm), are also sometimes referred to as light. X-rays, located in between UV and γ-rays, have a wavelength range of 10^{-2} Å to 10 nm. X-rays used in diffraction experiments have wavelengths of 0.5–2.5 Å because the interatomic distances in materials are on the order of 1–10 Å.

Electromagnetic Waves and X-rays

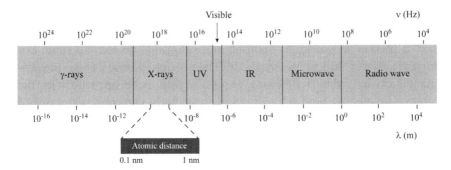

FIGURE 1.4 The spectrum of electromagnetic radiation.

Maxwell equations are a set of partial differential equations that form the foundation of classical electrodynamics, classical optics, and electric circuits. The fact that light is also an electromagnetic wave was revealed by Maxwell equations, which describe how electric and magnetic fields are generated and altered by each other. They are named after the Scottish physicist and mathematician James Clerk Maxwell. When Maxwell published his first extensive account of the electromagnetic theory in 1867, the frequency range was only known to extend from the IR, across the visible, to the UV. Although this range is of primary concern in optics, it is just a small segment of the whole electromagnetic spectrum. Ten years later, H. Hertz succeeded in producing and detecting electromagnetic waves. His transmitter was an oscillating discharge across a spark gap and he used an open loop of wire for a receiving antenna. A small spark induced between the two ends of the antenna indicated the detection of an incident electromagnetic wave. The waves used by Hertz are now classified in the radio frequency range, which extends from a few Hz to about 10^9 Hz (in wavelength, from several kilometers to 30 cm). These waves are usually generated by an assortment of electric circuits.

The microwave region extends from about 10^9 Hz to 3×10^{11} Hz and the corresponding wavelengths lies in between 0.3 m and 1 mm. Radiation that can penetrate the Earth's atmosphere ranges from less than 1 cm to about 30 m. Microwaves are thus of interest in space-vehicle communications. Molecules can absorb and emit energy by altering the state of motion of their constituent atoms. They can be made to vibrate and rotate. The energy associated with each motion is quantized. Namely, molecules possess vibrational and rotational energy bands as well as their electronic energy bands. Only polar molecules experience forces via the electric field of an incident electromagnetic wave. The polar molecules can absorb a photon and make a rotational transition into an excited state. For instance, water molecules are polar. Thus, when exposed to an electromagnetic wave, they will swing around, trying to be lined up with the alternating electric field. This effect is particularly vigorous at any one of their rotational resonances (i.e., rotational energy bands). Consequently, water molecules efficiently absorb microwave at or near such a resonance frequency. The microwave oven (12.2 cm, 2.45 GHz) is a well-known application, which utilizes heating by water molecules contained in the food. On the other hand, nonpolar molecules such as carbon dioxide, hydrogen, oxygen, and methane cannot make rotational transitions via the absorption of electromagnetic wave. Microwaves are also widely used for wireless communications.

The IR region extends approximately from 3×10^{11} Hz to 4×10^{14} Hz. The IR is often divided into four sub-regions: the near IR, i.e., near the visible (780–3000 nm), the intermediate IR (3000–6000 nm), the far IR (6,000–15,000 nm), and the extreme IR (15,000 nm–1 mm). This is a rather loose division, and there is again no sharp boundary separating them. Any material radiates and absorbs IR via thermal agitation of its constituents. Although the molecules of any objects at a temperature above $T = 0$ K will radiate IR, it is abundantly emitted in a continuous spectrum from hot bodies such as electric heaters, burning coals, and house radiators. Approximately half of the electromagnetic energy from the Sun is IR. The human body also radiates IR, even though the radiant energy is quite weak. A molecule can not only rotate but also vibrate in several different ways, with its atoms moving in various directions. For the vibration mode, the molecule need not be polar. For example, CO_2 has three vibrational modes and many associated energy levels, each of which can be excited by photons. The corresponding absorption spectra lie in the IR region. A number of molecules have both vibrational and rotational resonances and are good IR absorbers. We can feel the resulting build-up of thermal energy when our face was put in the sunshine. IR energy is usually measured by a device that responds to the heat generated on absorption. A small difference in the temperatures of an object and its surroundings results in characteristic IR emission, which can be effectively utilized for medical diagnostics.

Visible light has a very narrow band of frequencies from about 3.8×10^{14} Hz to 7.5×10^{14} Hz. It is usually generated by a rearrangement of the

Electromagnetic Waves and X-rays

outer electrons of atoms and molecules. The color of light is determined by its wavelength (and frequency). Newton was the first to recognize that white light is essentially a mixture of all the colors of the visible spectrum. Colors are actually the subjective human physiological responses to the various wavelength regions ranging from about 650 nm for red, through orange, yellow, green, and blue, to violet at about 400 nm. A variety of different wavelength mixtures can evoke the same color response from the eye-brain sensor. For example, a beam of red light overlapping a beam of green light will result in the sensing of yellow light, even though the overlapped beam has no wavelengths belonging to the yellow band. That is why a display can be operated with only three light sources: red, green, and blue. Next to visible light in the electromagnetic spectrum is the UV region that ranges approximately from 8×10^{14} Hz to about 3.4×10^{16} Hz. A UV photon can be emitted by an atom when its electron makes a long jump down from a highly excited state. Photon energies in UV range from roughly 3 eV to 100 eV. UV rays from the Sun thus have more than enough energy to ionize atoms. Fortunately, ozone in the atmosphere substantially absorbs a lethal UV stream from the Sun.

X-rays was fortuitously discovered by W. Röntgen in 1895. They have extremely short wavelengths; most are smaller than the atom size. The most practical method for producing X-rays is the rapid deceleration of charged particles accelerated to a very high speed. A broad X-ray spectrum arises when an energetic electron beam collides with a target material, such as a Cu plate. The atoms of the target may also be ionized during the bombardment. If the ionization occurs by removal of an inner electron tightly bound to the nucleus, the atom will emit X-rays as the vacant level is occupied by one of higher-lying electrons. The resulting quantized emissions are characteristic of the target atom, and accordingly are called *characteristic* radiation. γ-rays are the highest-photon energy, lowest-wavelength electromagnetic radiations. These rays are emitted by particles undergoing transitions within the atomic nucleus. Since the wavelengths are so short, it is very difficult to observe any wave-like properties from the γ-rays.

Why X-rays can penetrate deeply into metals is well explained by the concept of *plasma frequency*. Since the derivation of plasma frequency from the Maxwell equations can be found in many electromagnetics and optics books,[1-5] only its physical meaning is here mentioned. Free electrons under an electric field are accelerated due to the induced electrostatic force. At a fixed position, the direction of electric field of the electromagnetic wave changes with time. When an electromagnetic radiation is incident into a material, its free electrons will oscillate in response to the alternating electric field. The fact that the electrons are made to oscillate by the electromagnetic wave means that they absorb the radiation energy. Ordinary light is easily absorbed by free electrons within the metal and the absorbed energy is reflected as radiation or dissipated as heat. Since electrons also have a mass, they are unable to keep pace with the electric field oscillating at an extremely high rate. The plasma frequency means the maximum rate at which the free electrons of material can make a collective motion. The plasma frequencies of most metals lie in the UV region. Therefore, metals do not efficiently absorb the energy of X-rays because X-rays have a higher frequency than their plasma frequencies. That's why X-rays are permeable into metals. Of course, the penetration depth depends on the substance.

EXAMPLE 1.1.

Determine the wavelength, frequency, and speed of the wave functions $f_1 = \sin 2\pi(0.2x - 3t)$ and $f_2 = \sin(7x + 3.5t)$. The units of x and t are meters (m) and seconds (s).

Answer: From the general sinusoidal wave function

$$f = \sin\frac{2\pi}{\lambda}(x - \upsilon t) = \sin\left(\frac{2\pi}{\lambda}x - 2\pi\nu t\right)$$

we can know that f_1 has $\lambda = 5$ m, $\nu = 3$ Hz, and $\upsilon = 15$ m/s. f_2 is a wave propagating in the negative x-direction and its speed is $\upsilon = 0.5$ m/s. The wavelength and frequency of f_2 are $\lambda = 2\pi/7$ m and $\nu = 3.5/2\pi$ Hz.

1.3 WAVE-PARTICLE DUALITY

Electromagnetic radiation exhibits wave-like properties and particle-like properties at the same time (wave-particle duality). Therefore, the propagation of an electromagnetic wave can be considered as a stream of light particles called *photons*. According to quantum theory, the energy of a photon (E_p) is related to the frequency and wavelength of the wave as follows,

$$E_p = h\nu = h\frac{c}{\lambda} \quad (1.6)$$

where h is Plank constant (6.63×10^{-34} j·s). As a consequence of this duality, an electromagnetic wave with $\lambda = 100$ nm can be viewed as equivalent to a flow of photons with $E_p = 12.44$ eV (Figure 1.5). Various phenomena occurring with the electromagnetic radiations can be better explained sometimes with the wave concept, and sometimes, the particle concept. Wave characteristics are more apparent when the radiation is measured over relatively large time-scales and over large distances (e.g., interference and diffraction). On the contrary, particle characteristics will be more obvious in the case of absorption because it occurs quite fast in specific positions. The radiation intensity (I) means the energy transferred across unit cross-section per unit time and has units of j/m²·s.

FIGURE 1.5 Wave-particle duality.

When the wavelength and frequency are fixed, the intensity of the electromagnetic wave is proportional to the square of its amplitude. Thus, if the amplitude is doubled, the intensity becomes four times (Figure 1.6(a)). Viewed from the particle nature, this means that the number of photons crossing unit cross-section per unit time increases four-fold. Let's consider a case in which atoms or electrons are photoexcited to higher energy states by absorbing the incident electromagnetic wave (Figure 1.6(b)). When the amplitude of the wave is doubled, four times more photons will be absorbed as a result of the four-fold intensity increase. It is to be noted, however, that the energy of individual photons is fixed and the excitation to energy levels exceeding the photon energy is thus impossible no matter how high the intensity may be.

14   X-Ray Diffraction for Materials Research: From Fundamentals to Applications

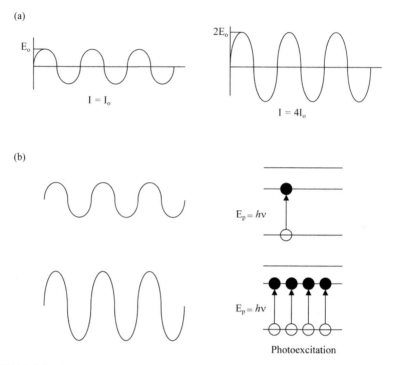

FIGURE 1.6 (a) Intensity proportional to the square of the amplitude. (b) A four-fold increase in the number of photons as a result of the doubled amplitude.

EXAMPLE 1.2.

If the intensity of an X-ray-beam at $\lambda = 10$ nm is four times higher than that of a visible light at 600 nm, how much different is the number of photons moving across unit area per unit time in two radiations?

Answer: Since the photon energy is inversely proportional to the wavelength as given by Eq. (1.6), the X-ray beam has a sixty times higher photon energy than the visible light. However, its intensity is just four times higher. Therefore, the visible light carries fifteen times more photons. Intensity means the energy transferred across unit area per unit time. Although individual X-ray photons have a much higher energy, the X-ray intensity is not so much high as its photon energy. Therefore, the visible light has more photons flowing per unit cross-section per unit time.

Electromagnetic Waves and X-rays

1.4 GENERATION OF X-RAYS

X-rays are generated when electrons accelerated to a very high speed rapidly decelerate. They are typically produced by an X-ray tube that contains two metal electrodes: a cathode (source of electrons) and an anode (metal target). The cathode is maintained at a high negative voltage (-V) and the anode, at ground potential. Figure 1.7 shows a schematic of the common filament-type tube. A tungsten filament within the cathode is heated by the passage of an electric current and produces electrons. A high electric potential in the range 20–60 kV accelerate the electrons emitted by the hot filament toward the metal target. When colliding with the target, the electrons lose their kinetic energy (KE) and the lost energy is emitted as X-rays. All these processes occur inside an evacuated glass envelope. X-rays are emitted from the anode in all possible directions. But only a narrow beam making a small angle with the target face is allowed to pass out of the evacuated tube through a window. The window is made of a substance with a very low absorption coefficient for X-rays; the substance for window is usually beryllium (Be). The KE of the electrons on impact is given by the following equation,

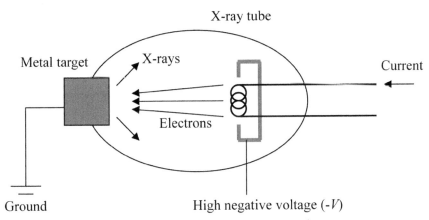

FIGURE 1.7 Schematic of the common filament-type X-ray tube.

$$KE = \frac{1}{2}mv^2 = eV \tag{1.7}$$

where m is the mass of the electron (9.11×10^{-31} kg) and v, its speed just before collision. The maximum energy of the emitted X-ray photons is limited by the energy of the incident electrons that is equal to the applied voltage times the electron charge. The production of X-rays is an inefficient process. Most of the electrical energy consumed by the tube is released as heat and approximately one percent is transformed into X-rays. Therefore, the X-ray tube should be designed to cool down the heated target. Usually, the heated target is cooled from behind by a flowing stream of water. The energy of the produced X-ray photons is maximized when the accelerated electrons completely stop in a single collision and transfer all their kinetic energy into the photon energy. The maximum X-ray frequency and minimum wavelength available in a given voltage is deduced from the following relation.

$$eV = hv_{max} = h\frac{c}{\lambda_{min}} \tag{1.8}$$

Equation(1.8) corresponds to an extreme case where the electron energy is 100% transformed into the photon energy. However, most electrons undergo multiple collisions and successively lose a part of their energy, emitting photons with energy less than hv_{max}. Thus, X-ray spectrum coming from the tube consists of many different wavelengths (and frequencies), which is known as *continuous radiation* or *white radiation*. The intensity of the emitted radiation will vary continuously with wavelength. The intensity at a fixed wavelength depends on the operating voltage of the tube and on the nature of the target metal. A typical X-ray spectrum from a molybdenum target is shown in Figure 1.8. As the applied potential difference increases, the *KE* of the electrons bombarding the target also increases. This will lead to an overall increase in the intensity of the emitted X-rays. The shortest wavelength, λ_{min}, is determined according to Eq. (1.8). The higher the accelerating voltage of the X-ray generator is, the shorter the minimum wavelength that can be obtained.

Electromagnetic Waves and X-rays

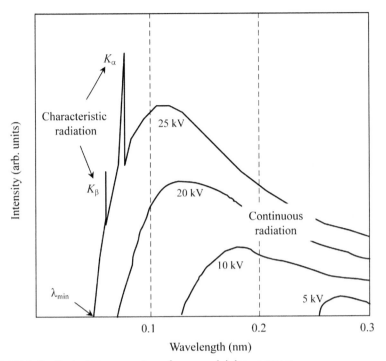

FIGURE 1.8 Typical X-ray spectrum from a molybdenum target.

When the energy of the accelerated electrons is higher than a threshold value (which depends on the metal target), sharp peaks appear at certain wavelengths superimposed on the white radiation. They are called *characteristic lines* and the wavelengths of these peaks depend solely on the target material. While the continuous radiation is caused by the KE loss of electrons in a series of collisions, the characteristic radiation arises from the ejection of an electron from one of the inner shells of the target atom. If one of the electrons striking the target has sufficiently high energy, it can knock an electron out of the K shell of the target atom, leaving the atom in an excited state. This will result in the transfer of an electron from outer shells to the vacant K level in order to lower the overall energy. Such a transfer will be followed by the emission of an X-ray photon whose energy is equal to the difference in energy between the two different states. The radiation emitted as a result of such a process will thus have a definite wavelength characteristic of the target element. The characteristic lines referred to as K, L, M, etc., correspond to transitions to the K, L, M shells, respectively. The vacant K level may be filled by an electron from any of the outer shells. When the two orbitals involved in the transition are adjacent, the line is represented by a subscript α. If the involved orbitals are separated by two levels, the line is designated as β. For instance, when an electron is ejected from the K shell and its vacant site is occupied by an electron from the L shell, the K_α line is emitted (Figure 1.9). The K_β transition refers to the case where the vacant K shell is filled by an electron from the M shell. Since the β-transition has a larger energy difference than the α-transition, the K_β line exhibits higher photon energy (i.e., shorter wavelength) than the K_α line. However, the K_α line is much stronger than the K_β line because the vacant site of the K shell is more probably occupied by an L electron than by an M electron. When the L shell becomes vacant as a result of the K_α transition, the vacancy will also be filled by an electron from the outer shells. Therefore, the K_α line is always accompanied by the L transition.

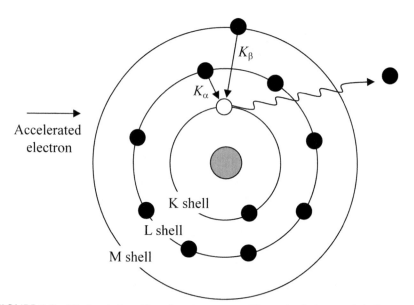

FIGURE 1.9 Electronic transitions in a target atom. The emitted characteristic lines are represented by arrows.

Only the characteristic lines are used in most X-ray diffraction experiments, except the Laue method that requires a white X-ray spectrum. The L lines are not suitable for X-ray diffraction because they have rather long

Electromagnetic Waves and X-rays

wavelengths and also are substantially absorbed by the target before escaping it. The characteristic lines of common target metals are listed in Table 1.1. All target elements give rise to two K_α lines: $K_{\alpha1}$ and $K_{\alpha2}$. Since the two lines are very close in wavelength, they are not always resolved as separate lines. The $K_{\alpha1}$ line has the shorter wavelength and is about twice as strong as the $K_{\alpha2}$ line. If the two lines are not resolved, the doublet is simply called the K_α line. The wavelength of an unresolved K_α doublet is usually given by the weighted average of the wavelengths of its components. It is written as $\lambda_{K\alpha} = \frac{2}{3}\lambda_{K\alpha1} + \frac{1}{3}\lambda_{K\alpha2}$. Thus, the wavelength of the unresolved Cu K_α line is $(2\times1.541 + 1.544)/3 = 1.542$ Å. The Cu K_α line is widely employed in many X-ray diffractometers. While the wavelength of characteristic radiation is dependent only on the target element, its intensity is influenced by the potential applied across the tube. If the applied potential is below a certain threshold value, none of the accelerated electrons will have sufficient energy to eject an electron from the target atom. The operation voltage of an X-ray tube should therefore be greater than this threshold value. It is maintained so that the characteristic radiation has an optimal intensity with respect to the white radiation.

TABLE 1.1 Wavelengths of the Characteristic Lines of Common Target Metals

Target	Cu		Mo		Fe		Co		Cr	
Lines	K_α	K_β	K_α	K_β	K_α	K_β	K_α	K_β	K_α	K_β
λ (Å)	1.542	1.392	0.711	0.632	1.937	1.757	1.790	1.621	2.291	2.085

A specific characteristic line can be selected by passing the output X-ray beam from the tube through a filter made of a material whose absorption edge lies in between the K_α and K_β wavelengths. For instance, a Ni filter absorbs the Cu K_β line much strongly than the Cu K_α line. After passing through the filter, the K_β line is reduced to negligible intensity, while the intensity of the K_α line decreases only by a small factor. Although the primary purpose of such a filter is to remove the K_β line, white radiation whose wavelength is below the absorption edge of the filter will also be cut out. Since the filtered output beam can have a fairly intense K_α line superimposed on the weak continuous spectrum, it may be sufficient merely to remove the K_β line for some applications. However, many X-ray diffraction experiments require monochromatic radiation that has a fixed wavelength (and frequency). Even though the characteristic peaks have a very narrow wavelength range, they are not perfectly monochromatic. Modern diffractometers are equipped with a single-crystal monochromator to make the characteristic beam as close to monochromatic as possible. The monochromator is aligned such that only a central wavelength component is diffracted by the high-quality crystal. This procedure enables the characteristic peak to have a much narrower wavelength range than before, although its intensity is inevitably reduced to a degree. A characteristic beam diffracted by the monochromator is used as the monochromatic X-ray source in many experiments.

1.5 ABSORPTION

We now take a look at the absorption of electromagnetic radiation. Absorption is a physical phenomenon in which the energy of a photon is taken up by matter, typically by the electrons of an atom. The energy absorbed by matter is released in the form of radiation or simply dissipated as heat. The intensity of an electromagnetic wave decreases as it propagates through an absorptive medium. In general, the degree to which absorption occurs is not influenced by the intensity (linear absorption), although the medium can change its transparency depending on the intensity (nonlinear effect). In the general case, the fractional decrease in intensity is directly proportional to the distance by which the wave travels. In differential form,

$$\frac{-dI}{I} = \alpha dx \tag{1.9}$$

where the proportionality constant α is called the *linear absorption coefficient*. The absorption coefficient depends on the wavelength of the electromagnetic wave as well as the medium. Integration of Eq. (1.9) gives

$$I(x) = I_o e^{-\alpha x} \tag{1.10}$$

where I_0 is the initial intensity and $I(x)$ is the intensity after traveling by a distance x. The absorption coefficient of a material is usually obtained by measuring the transmittance (T). The transmittance is defined as the ratio of the transmitted intensity to the incident intensity, i.e., $T = I/I_i$. At least two samples of different thickness are needed to precisely measure the absorption coefficient of the material. Suppose that an electromagnetic wave with I_i is incident into a medium of thickness d_1 (Figure 1.10). The transmittance of the medium will be given by $T_1 = I_t/I_i = Ae^{-\alpha d_1}$, where A is a constant reflecting the intensity loss due to reflection at the interfaces. It is impossible to derive the absorption coefficient from T_1 alone if the unknown A value is much deviating from unity. The transmittance of another sample of thickness d_2 is expressed as $T_2 = I_t/I_i = Ae^{-\alpha d_2}$. Since the reflection loss is independent of the thickness, the absorption coefficient can be accurately determined from the following relation.

$$\alpha = \frac{\ln(T_1/T_2)}{d_2 - d_1} \tag{1.11}$$

As shown in Eq. (1.10), the intensity inside a material decays exponentially with distance from its surface. *Penetration depth* is defined as the depth at which the intensity falls to 1/e of its value at (more precisely just beneath) the surface. It is a measure of how deeply an electromagnetic radiation can penetrate into the material and is equal to the inverse of the absorption coefficient. The penetration depth also varies with the substance and the radiation wavelength. For instance, the penetration depth of visible light into metals ranges from a few nm to tens of nm, depending on their electrical conductivity.

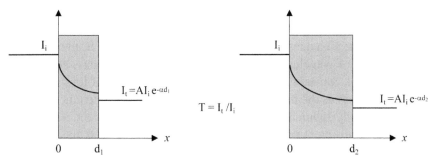

FIGURE 1.10 Measurement of the absorption coefficient. The absorption coefficient of a material can be precisely determined by measuring the transmittances of two samples with different thicknesses.

EXAMPLE 1.3.

A certain material has an absorption coefficient $\alpha = 0.2$ cm^{-1} at 500 nm. If a sample of 1 mm thickness exhibits a transmittance of 0.7 at this wavelength, what will the transmittance of a 3 mm-thick sample be?

Answer: Since the 1 mm-sample exhibits $T = 0.7$, we obtain the relation of $0.7 = Ae^{-0.2\times0.1}$ from which the reflection loss, represented by the constant A, is known. The transmittance of the 3 mm-thick sample will be given by $T = Ae^{-0.2\times0.3}$. By comparison of the measured and expected transmittances, we know that the 3 mm-thick sample will have a transmittance of

$$T = 0.7e^{-(0.06-0.02)} = 0.673$$

The linear absorption coefficient is dependent on the material considered, its density, and the wavelength of the incident radiation. Since the linear absorption coefficient α of a material is proportional to its density ρ, absorption is often described with the *mass absorption coefficient* defined as α/ρ. Equation(1.10) may be rewritten in terms of the mass absorption coefficient.

$$I(x) = I_o e^{-(\alpha/\rho)\rho x} \tag{1.12}$$

where ρx is the area density also known as mass thickness. Both α and α/ρ are functions of wavelength. At a given wavelength, the mass absorption coefficient is a constant of the material and independent of its state (solid or liquid), whereas the linear absorption coefficient is a state function. The quantity usually tabulated is the mass absorption coefficient. The linear absorption coefficient may be calculated from the tabulated mass absorption coefficients of the elements (International Tables for Crystallography, Vol. 3),[6] provided the composition and density of the material are known. The mass absorption coefficient is a convenient concept especially when we deal with the absorption of a substance containing more than one element. Regardless of whether the substance is a mechanical mixture or a solution, its mass absorption coefficient is simply the weighted average of the mass absorption coefficients of its constituent elements. Consider a

Electromagnetic Waves and X-rays

solid solution containing three different elements. If w_1, w_2, and w_3 are the weight fractions of these elements in the solution and $(\alpha/\rho)_1$, $(\alpha/\rho)_2$, and $(\alpha/\rho)_3$ are their mass absorption coefficients, the mass absorption coefficient of this solution is then given by

$$\alpha/\rho = w_1(\alpha/\rho)_1 + w_2(\alpha/\rho)_2 + w_3(\alpha/\rho)_3 \tag{1.13}$$

This is based on the fact that the mass absorption coefficient of an element is independent of its concentration in the solution. Eq. (1.13) can be alternatively expressed as

$$\alpha = \rho_1(\alpha/\rho)_1 + \rho_2(\alpha/\rho)_2 + \rho_3(\alpha/\rho)_3 \tag{1.14}$$

where ρ_1, ρ_2, and ρ_3 are the densities of constituent elements of the solution. We can obtain the linear absorption coefficient of a solution substance if the densities of its constituent elements are known.

The absorption of X-rays by a matter takes place in two distinct ways: scattering and true absorption. These two processes constitute the total absorption characterized by the quantity α/ρ. The scattering of X-rays by atoms may also occur in two different ways, both of which involve interaction between X-radiation and electrons. An X-ray photon encountered with a loosely bound or free electron can be deflected by the electromagnetic field of the electron, giving some of its energy to the electron as kinetic energy. Thus, the deflected (i.e., scattered) X-ray photon has lower energy and longer wavelength than the incident photon. This *incoherent* scattering, discovered by A. Compton, is called the *Compton scattering* or *effect*. The Compton scattering can be understood only by considering the incident beam as a stream of X-ray photons (quanta). It may be viewed as a collision between two billiard balls. The effect is of scientific significance because it demonstrates that light cannot be explained purely as a wave phenomenon. Compton derived the mathematical relationship between the scattering angle of the X-rays and the shift in wavelength by assuming that each scattered X-ray photon interacted with only one electron. It is important to note that the phase of the Compton-scattered radiation has no fixed relation to the phase of the incident beam. This incoherent radiation cannot participate in X-ray diffraction because interference does not take place between the waves at random phases.

Another way in which X-rays are scattered by atoms can be explained by treating the incident X-ray beam as a wave with oscillating electric field. When the incident wave encounters an electron, its time-varying electric field will cause the electron to oscillate about a mean position. An oscillating charge emits an electromagnetic wave. The oscillating electron radiates X-rays of the same wavelength and frequency as the incident beam. In this *coherent* scattering, there is a definite phase relationship between the scattered beam and the incident beam. The interference associated with such coherent scattering is the basis of X-ray diffraction and will be discussed in Chapters 3 and 5. Since the phase change on coherent scattering is identical for all the electrons in a material, we do not have to consider it in deriving the condition for diffraction. When X-rays in the range 0.5–2.5 Å are incident on crystalline materials, diffraction patterns are observable because the distances between adjacent atoms are on the same scale and because the incident X-rays are coherently scattered from the electrons of the constituent atoms. Incoherent scattering occurs simultaneously, which contributes only to the background of the diffraction pattern.

True absorption arises from electronic transitions within the atom. We already know that an electron of sufficient energy can knock a K electron out of an atom. Likewise, an electron can be ejected from the atom by an X-ray photon of the corresponding energy. Since the vacant K shell will be filled by an electron from outer shells, this photo-absorption process will be accompanied by the emission of characteristic K radiation. The emitted characteristic radiation is called *fluorescent radiation*. Such fluorescent X-rays, radiated in all directions, may be reabsorbed by another atom. The way in which the absorption coefficient varies with wavelength is helpful to figure out the interaction of X-rays and atoms. The mass absorption coefficient of Ni is shown in Figure 1.11; it is typical of all materials. The overall absorption drastically decreases with decreasing wavelength. This reflects the fact that the scattering-induced energy loss is approximately proportional to λ^3 in the X-ray wavelength range. There is a sharp discontinuity (a jump) in the absorption curve, called an *absorption edge*. This absorption edge is generated because the energy of an X-ray photon is absorbed to eject a K electron out of a Ni atom. The K absorption edge of Ni is located at 1.49 Å. Thus, X-rays at wavelengths slightly shorter than this value can be significantly absorbed by Ni. For this reason, a Ni filter is used to cut out the Cu K_β line at 1.39 Å.

Electromagnetic Waves and X-rays

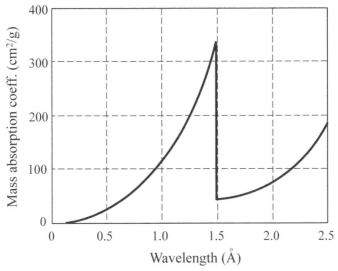

FIGURE 1.11 Mass absorption coefficient of Ni at X-ray wavelengths.

Problems

1.1. Explain why X-rays are useful for the structural analysis of materials, while visible or infrared lights are not.

1.2. Show that $f(x-vt)$ is a solution of the following one-dimensional wave equation

$$\frac{\partial^2 f}{\partial t^2} = v^2 \frac{\partial^2 f}{\partial x^2}$$

where f is any differentiable function with the argument $x-vt$.

1.3. Using the wave functions $f_1 = 4\sin 2\pi(0.2x - 3t)$ and $f_2 = 0.2\sin(7x + 3.5t)$, determine in each case the values of (a) frequency, (b) wavelength, (c) period, (d) amplitude, and (e) direction of motion. t is in seconds and x is in meters.

1.4. When an X-ray tube is operated at 60 kV, what are the kinetic energy and speed with which the electron strikes the metal target? Calculate the minimum wavelength and the maximum photon energy achievable under this condition.

1.5. Characteristic K_α line is always accompanied by L line, but the L line is seldom used for X-ray diffraction. Explain why?

1.6. When an X-ray beam is incident into a crystalline matter, incoherent and coherent scattering takes place simultaneously as a consequence of the interaction with electrons. However, only the coherently scattered X-rays are involved in diffraction. Why?

1.7. The penetration depth of visible light into metals ranges from a few nm to tens of nm. If the penetration depth of a certain metal is 12 nm at $\lambda = 632.8$ nm, how thin should it be to exhibit 50% transmittance at the same wavelength?

1.8. An X-ray beam of $\lambda = 1.54$ Å is propagating in free space. If this beam has an intensity of 100 W/cm^2, then how many photons are moving across unit area per unit time?

CHAPTER 2

GEOMETRY OF CRYSTALS

CONTENTS

2.1   Introduction 28
2.2   Lattice 29
2.3   Crystal Systems 31
2.4   Directions and Planes 45
2.5   Reciprocal Lattice 51
2.6   Crystal Structures 66
2.7   Stereographic Projection 86
Problems 89

2.1 INTRODUCTION

A crystal is a solid material whose constituent atoms are periodically arranged in three dimensions. Crystallography refers to the scientific area that studies the arrangement of atoms in solids. Not all materials are crystalline and some are amorphous. With respect to the atomic arrangement, there is no fundamental difference between an amorphous solid and a liquid. In this sense, the former is often regarded as a "super-cooled liquid". Most solids are composed of regularly arranged atoms because the crystalline state is energetically more stable. The regularity of atomic arrangement can be described by symmetry elements, which ultimately determines the physical properties of a crystal. Crystallography is a very broad subject beyond simply comprehending the crystal structure. In-depth knowledge of crystallography is central to the study of many active areas in materials science, chemistry, earth science, and physics. When performing any process on a material, it may be required to find out what phases are present in the material. Each phase has a characteristic arrangement of atoms and symmetry. Crystallography covers a huge number of symmetry patterns that can be formed by atoms in a crystal and has a relation to group theory (point group and space group). In this chapter, we explain the basic aspects of crystallography that are necessary for the understanding of X-ray diffraction and its applications. These include the concept of lattice and reciprocal lattice, symmetry elements, crystal systems, and the crystal structures of some important materials. How the interplanar distances and angles in crystals can be determined are also described. Textbooks by Kelly/Knowles[7] and McKie/McKie[8] would be greatly helpful to acquire deeper knowledge on the crystallography.

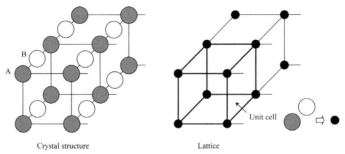

FIGURE 2.1 Crystal structure and lattice. Lattice is a three-dimensional array of lattice points that have identical surroundings.

2.2 LATTICE

In this chapter, we explain the principle of geometrical crystallography by considering perfect single crystals. The regular arrangement of atoms in a single crystal can be completely described by defining a fundamental *repeat unit* coupled with a statement of the translations necessary to build the crystal from the repeat unit. A *space lattice* (simply called a *lattice*) is an array of hypothetical points introduced to figure out how the atoms of a crystal are periodically arranged in space. Each of the points, called a *lattice point*, has identical surroundings and the lattice corresponds to a three-dimensional network of lattice points. Let's consider a crystal consisting of A and B atoms, as shown in Figure 2.1. The surroundings of the A atom are different from those of the B atom. If a pair of A and B atoms are allocated to one lattice point, all lattice points have identical surroundings and are indistinguishable from one another. These two atoms are called the *basis* of the lattice. The crystal structure is constructed by placing the basis atoms on each lattice point. The lattice then represents an essential element of the translational symmetry of the crystal. The space lattice can be built up by repetition of a repeat unit, known as the *unit cell*. A general form of the unit cell is a parallelepiped that contains just one lattice point (Figure 2.2). The sides of the unit cell are taken as the axes of the crystal. The size and shape of the unit cell are described with three vectors **a**, **b**, and **c** drawn from one corner of the cell. The x, y, and z directions of a crystal and its unit cell are taken parallel to these unit cell vectors. The lengths of **a**, **b**, and **c** vectors are denoted by a, b, and c, and the angles between them, α, β, and γ. These values are called the *lattice parameters* or *constants* of the crystal.

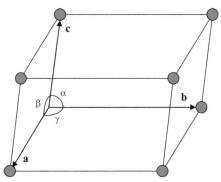

FIGURE 2.2 A general form of the unit cell. The lengths of **a**, **b**, **c** vectors are a, b, c, respectively, and the interaxial angles are denoted by α, β, and γ.

There are various ways to choose a unit cell. The smallest repeat unit in space lattice is referred to as the *primitive cell* and it contains only one lattice point. However, it is more convenient to consider a larger repeat unit in many cases. This *non-primitive cell* contains more than one lattice point. A cubic cell is conventionally taken as the unit cell of the face-centered cubic lattice, while its primitive cell is rhombohedral (Figure 2.3). The axes of the conventional cubic cell have an equal length and are at right angles with one another, i.e., $a = b = c$ and $\alpha = \beta = \gamma = 90°$. Thus, the periodicity and symmetry of the lattice can be more easily visualized with a cubic cell. The inter-axial angle of the rhombohedral cell is 60° and its axis length is $1/\sqrt{2}$ times the length of the cubic axis. Since each cubic cell has four lattice points, its volume is also four times that of the primitive cell.

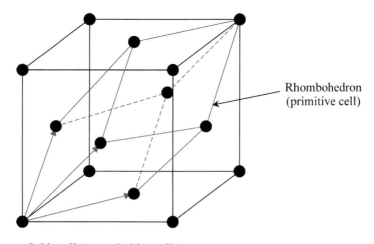

FIGURE 2.3 The relationship between the primitive cell and the non-primitive cubic cell in the face-centered cubic lattice.

There are four different lattice types: simple, body-centered, base-centered, and face-centered (Figure 2.4). When the lattice points are located only on the corners of the unit cell, it is called *simple lattice*. Thus, the lattice type of all the primitive cells is simple. The *body-centered lattice* has another lattice point on the body center of the cell, and the *base-centered lattice*, two more lattice points on the centers of two parallel faces. In the *face-centered lattice*, all of the corners and face centers of the unit cell are

Geometry of Crystals

occupied with lattice points. The number of lattice points per unit cell of each type is one, two, two, and four, respectively. The crystal structure of a material is described by the arrangement of atoms within its unit cell. The positions of the atoms inside the unit cell are given by the set of atomic positions measured from a lattice point. This lattice point, arbitrarily chosen, becomes the origin of the unit cell. Commonly, atomic positions are represented in terms of factional coordinates, relative to the unit cell lengths. For instance, an atom located at the center of a unit cell has atomic position (1/2,1/2,1/2), regardless of the shape and size of the unit cell.

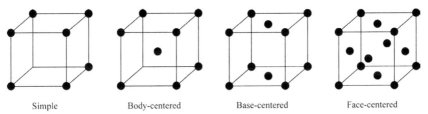

FIGURE 2.4 Four different lattice types that can be possessed by crystals.

2.3 CRYSTAL SYSTEMS

2.3.1 SYMMETRY

In the previous section, the regular arrangement of atoms in a crystalline solid was described with respect to the concept of a lattice. The lattice is a regular array of hypothetical points in which each lattice point has identical surroundings in the same direction. As the unit cell is a repeat unit, the space lattice is built up by stacking unit cells in three dimensions. The crystal structure can be completely described by stating the lattice constants (i.e., the unit cell dimensions) and the coordinates of atoms within the unit cell. Any two atoms separated by a lattice translation should be equivalent in every aspect; they should be of the same element and also have identical surroundings in the same direction. Different from such lattice repetition, there is another kind of repetition known as *symmetry*. The symmetry arises when an atom or group of atoms is regularly repeated to a pattern. The symmetric arrangement of atoms in a crystal is described in terms of the symmetry elements. There are four pure symmetry elements or operators: *rotation, reflection, inversion,* and *rotation-inversion.*

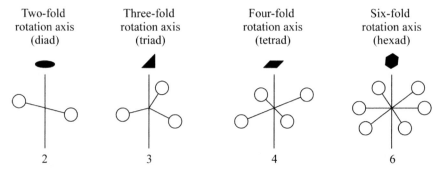

FIGURE 2.5 Numerical and graphical symbols for the rotational symmetry axes. Objects come into self-coincidence after rotation of 360°/n about the symmetry axis. The possible values of n are 1, 2, 3, 4, and 6.

A body is said to possess n-fold rotational symmetry if it comes into self-coincidence after rotation of 360°/n about an axis. Axes of rotational symmetry include one-fold, two-fold, three-fold, four-fold, and six-fold. These correspond to repetition every 360°, 180°, 120°, 90°, and 60° and are called monad, diad, triad, tetrad, and hexad, respectively. Although the possible values of n are 1, 2, 3, 4, and 6, a one-fold axis, which brings the crystal into self-coincidence after rotation of 360°, is obviously trivial. A five-fold axis or one of higher degree than six is impossible because such symmetry does not fill up space without gaps. Just as there are no pentagon-shaped paving blocks in the sidewalk, five-fold rotational symmetry cannot exist in a single crystal. Crystals can have diad, triad, tetrad, and hexad only. These rotation axes are denoted by the numerical symbols 2, 3, 4, and 6. The graphical symbol for each axis is also shown in Figure 2.5. Another type of symmetry is reflection. The operation is that of reflection in a mirror. The mirror plane marked m in Figure 2.6 runs normal to the page and reflects a right-handed object to a left-handed one and vice versa. It should be noted that the right-handed object cannot be translated in the plane of the page to superimpose on the left-handed one. Rotation and reflection are easily combined to give higher symmetry. Figure 2.7(a) shows a two-dimensional pattern consisting of circles with different sizes. This pattern possesses a tetrad axis and mirror planes running normal to the

Geometry of Crystals

page. However, once the small-circle pairs are rotated like Figure 2.7(b), the mirror symmetry no longer exists and only a tetrad remains.

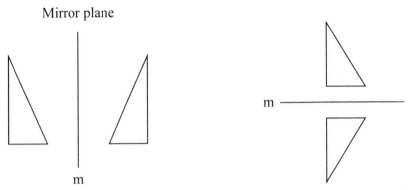

FIGURE 2.6 Mirror plane reflects a right-handed object to a left-handed one and vice versa.

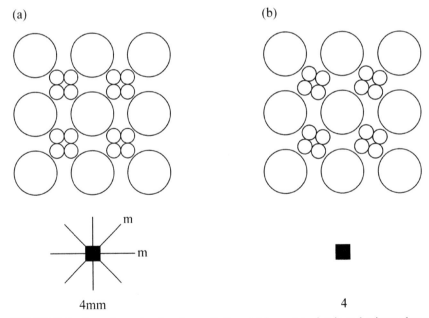

FIGURE 2.7 Two-dimensional patterns. Pattern (a) has a tetrad axis and mirror planes running normal to the page. Pattern (b) lacks mirror symmetry.

A center of symmetry is a point through which inversion produces an identical arrangement. When an atom with coordinates (x, y, z) in a crystal is duplicated by another atom of the same element with coordinates $(-x, -y, -z)$, the structure is said to possess an inversion center, i.e., a center of symmetry at the origin $(0, 0, 0)$. For instance, if we stand at the inversion center and look in a certain direction, we will find an identical outlook when we look in the opposite direction. Any crystals that exhibit inversion symmetry have a center of symmetry. The center of symmetry is alternatively called the point of inversion. The crystal needs to possess a center of symmetry for some applications, but it should be absent for others (see Example 2.3). The pure rotation axes designated as 2, 3, 4, and 6 rotate an object through $360°/n$ and have already been discussed. Rotation-inversion refers to the combination of two symmetry elements. This involves a rotation through $360°/n$, followed by an inversion through a center of symmetry on the axis of rotation. The basic operations of repetition by rotation axes are shown in Figure 2.8, together with their stereographic diagrams. An n-fold axis repeats an object by successive rotations through an angle of $360°/n$. Since the operation of one-fold rotation is self-coincident, any line passing through a center of symmetry can be a one-fold inversion axis. The operation of the other rotation-inversion axes is given in Figure 2.9. The two-fold rotation-inversion axis, denoted by $\bar{2}$, repeats an object by rotation through $360°/2 = 180°$ to give the dotted object, followed by inversion to give the full object. The objects located above and below the center of symmetry are represented as closed and open circles on the stereogram, respectively. Similarly, the three-fold rotation-inversion axis $\bar{3}$ involves a rotation through $120°$ coupled with an inversion. It is to be noted that the rotation and inversion are both part of the whole operation and should not be regarded as separate operations. The graphical symbols for the rotation-inversion axes are also shown in Figure 2.9.

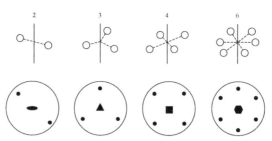

FIGURE 2.8 Basic operations of repetition by rotation axes along with their stereographic diagrams.

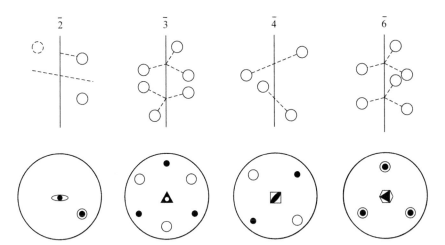

FIGURE 2.9 Operations of rotation-inversion symmetry axes. Their graphical symbols are given on the diagrams.

2.3.2 CRYSTAL SYSTEMS

Crystals are classified into seven crystal systems according to their rotational symmetry elements (including rotation-inversion symmetry). These are *triclinic, monoclinic, orthorhombic, trigonal, tetragonal, hexagonal,* and *cubic*. The trigonal system is sometimes regarded as a subdivision of the hexagonal system, rendering the number of systems to six. The minimum symmetry elements required for each system are listed in Table 2.1 and are also schematically illustrated in Figure 2.10. One system is distinguished from another by its symmetry elements. The presence of a certain minimum set of symmetry elements is an intrinsic property of each system. The classification of the crystal systems is purely based on the minimum requirement. Therefore, crystals may possess more than the minimum symmetry elements imposed by the system to which they belong. While the minimum requirement of the tetragonal system is one four-fold rotation (or rotation-inversion) axis, many tetragonal crystals have mirror planes running perpendicular and parallel to a tetrad. The seven crystal systems are characterized as follows.

TABLE 2.1 Symmetry Elements of the Crystal Systems

System	Minimum symmetry elements
Triclinic	No rotation symmetry
Monoclinic	One two-fold rotation (or rotation-inversion) axis
Orthorhombic	Three perpendicular two-fold rotation (or rotation-inversion) axes
Trigonal	One three-fold rotation (or rotation-inversion) axis
Hexagonal	One six-fold rotation (or rotation-inversion) axis
Tetragonal	One four-fold rotation (or rotation-inversion) axis
Cubic	Four three-fold rotation (or rotation-inversion) axes

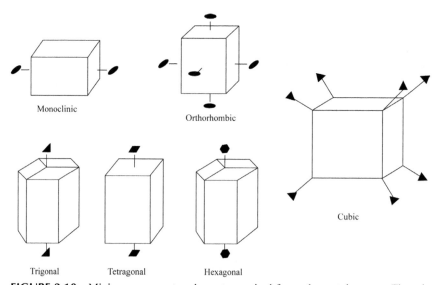

FIGURE 2.10 Minimum symmetry elements required for each crystal system. There is no requirement for the triclinic system.

Triclinic: The triclinic system has no rotation symmetry. The only symmetry element that a triclinic crystal can have is a one-fold rotation or rotation-inversion axis. This places no restriction on the shape of the unit cell. The unit cell of the triclinic system is a general parallelepiped with $a \neq b \neq c$, $\alpha \neq \beta \neq \gamma$. The symbol $\neq$ means "not necessarily equal to".

Monoclinic: The characteristic symmetry of the monoclinic system is a two-fold rotation (or rotation-inversion) axis, which is usually taken

Geometry of Crystals

along the y axis of the unit cell. The positive x and z axes are conventionally chosen so that the angle between them, β, is obtuse. The unit cell geometry is thus $a \neq b \neq c$, $\alpha = \gamma = 90°$, $\beta \geq 90°$.

Orthorhombic: Crystals in this system possesses three mutually perpendicular two-fold rotation (or rotation-inversion) axes. It is obviously convenient to take the x, y, and z axes parallel to the symmetry axes so that the unit cell becomes a rectangular parallelepiped. The sides of the unit cell are in general unequal to one another and the geometry is given by $a \neq b \neq c$, $\alpha = \beta = \gamma = 90°$.

Trigonal: This crystal system is characterized by the presence of a three-fold rotation (or rotation-inversion) axis. The unit cell is a rhombohedron with $a = b = c$, $\alpha = \beta = \gamma \neq 90°$. Although the x, y, and z axes of the unit cell are equally inclined to the triad, none of them are parallel to the symmetry axis. For this reason, a triple hexagonal cell is more frequently used for the description of trigonal crystals than the primitive rhombohedral cell. This is simply for convenience and is discussed in the next section.

Tetragonal: The tetragonal system has a four-fold rotation (or rotation-inversion) axis along the z axis of the unit cell whose geometry is given by $a = b \neq c$, $\alpha = \beta = \gamma = 90°$.

Hexagonal: The hexagonal system is characterized by the presence of a six-fold rotation (or rotation-inversion) axis, which is taken along the z axis of the unit cell. The x and y axes of the unit cell are at $120°$ and perpendicular to the symmetry axis. The hexagonal unit cell thus has $a = b \neq c$, $\alpha = \beta = 90°$, $\gamma = 120°$.

Cubic: This system possesses four three-fold rotation (or rotation-inversion) axes along the body diagonals of the unit cell, which is a cube. "Cubic" in the crystal system is a term different from the adjective of cube, i.e., regular hexahedron. A crystal without four triads is not cubic, even though it may have a unit cell with $a = b = c$, $\alpha = \beta = \gamma = 90°$. The cubic system is characterized by triads equally inclined to the three orthogonal reference axes x, y, and z.

The crystal system is determined by the symmetry elements that a crystal possesses, not by its lattice constants. Let's consider two hypothetic crystal structures as shown in Figure 2.11. The unit cells of both crystals have the same dimensions. One is surely cubic because four triads exist along the body diagonals of the unit cell. However, the other is tetragonal due to the absence of the three-fold rotational symmetry. The cubic crystal comes into self-coincidence when rotated through 120° about any of the triads, making the *x*, *y*, and *z* directions of the unit cell indistinguishable from one another. The unit cell dimensions of $a = b = c$ and $\alpha = \beta = \gamma = 90°$ are a consequence of the presence of four triads. If the crystal is cubic, its lattice parameters definitely have this relation. However, the opposite is not always true. Trigonal, tetragonal, and hexagonal crystals have a single characteristic symmetry axis. These crystals are thus referred to as uniaxial crystals. The macroscopic properties of uniaxial crystals, for example, electrical conduction, thermal expansion, and optical refraction, are anisotropic. The properties occurring along the characteristic axis are usually different from those measured in a direction normal to it.

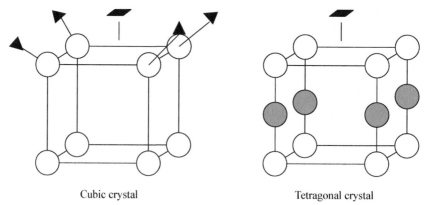

FIGURE 2.11 Two different crystal structures having the same unit cell dimensions. One is cubic and the other one is tetragonal.

2.3.3 BRAVAIS LATTICES

Seven different space lattices can be obtained by simply putting lattice points on the corners of the unit cells of the seven crystal systems. However, there are other ways to build up lattice points in space while maintaining the symmetry requirements imposed on such systems. The French scientist Bravais demonstrated that there are fourteen possible lattices and no more. These fourteen lattices are called the *Bravais lattices* after him. Table 2.2 lists the lattice types that each crystal system can have. For instance, orthorhombic crystals can exhibit four different types of lattices: simple, base-centered, body-centered, and face-centered. On the contrary,

Geometry of Crystals

only simple lattice is allowed for the trigonal and hexagonal systems. The fourteen Bravais lattices can be derived from two-dimensional lattices. All the fourteen three-dimensional lattices are derived in Appendix. A few examples are presented here. Figure 2.12(a) shows a two-dimensional lattice of tri-equiangular shape. This lattice possesses six-fold and three-fold rotation axes (also two-fold axes) perpendicular to the page. Space lattices are constructed by stacking tri-equiangular nets in a sequence. Six-fold axes exist only at the lattice points of the net. To preserve six-fold rotational symmetry in a space lattice, the nets should be stacked vertically above one another so that the lattice points in a net overlap those of other nets when viewed along the symmetry axis. The constructed space lattice has a unit cell as shown in Figure 2.12(b). This is a simple hexagonal lattice with $a = b \neq c$, $\alpha = \beta = 90°$, $\gamma = 120°$.

TABLE 2.2 Crystal Systems and Bravais Lattices

Crystal systems	Unit cell dimensions	Bravais lattice
Triclinic	$a \neq b \neq c$, $\alpha \neq \beta \neq \gamma$	Simple
Monoclinic	$a \neq b \neq c$, $\alpha = \gamma = 90°$, $\beta \geq 90°$	Simple Base-centered
Orthorhombic	$a \neq b \neq c$, $\alpha = \beta = \gamma = 90°$	Simple Base-centered Body-centered Face-centered
Trigonal	$a = b = c$, $\alpha = \beta = \gamma \neq 90°$	Simple
Hexagonal	$a = b \neq c$, $\alpha = \beta = 90°$, $\gamma = 120°$	Simple
Tetragonal	$a = b \neq c$, $\alpha = \beta = \gamma = 90°$	Simple Body-centered
Cubic	$a = b = c$, $\alpha = \beta = \gamma = 90°$	Simple Body-centered Face-centered

*The symbol $\neq$ means "not necessarily equal to".

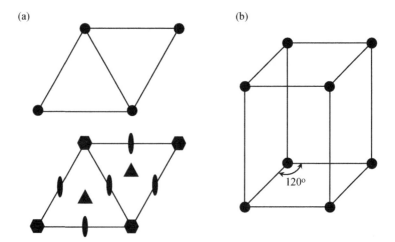

FIGURE 2.12 (a) Two-dimensional lattice of tri-equiangular shape and its rotational symmetry axes. (b) A space lattice constructed by stacking tri-equiangular nets such that six-fold rotational symmetry is maintained.

A space lattice consistent with three-fold rotational symmetry can be obtained by stacking tri-equiangular layers in a staggered fashion. As shown in Figure 2.12(a), the three-fold rotation axes run through the centers of triangles formed by the lattice points. Thus, if the second layer is stacked in such a way that its lattice points are placed above the centers of either upright or inverted triangles of the first layer, the three-fold symmetry is maintained (Figure 2.13(a)). To meet the fundamental translational symmetry (i.e., periodicity) of crystals, the third layer should be stacked by the same fashion. If stacking proceeds in this way, the fourth layer overlaps the first layer when viewed vertically. In Figure 2.13(a), the lattice points of the second and third layers were represented as gray and white dots, respectively and those of the first and fourth layers, black dots. The primitive cell of the constructed lattice is a rhombohedron as depicted in Figure 2.13(b). The edges of the cell are of equal length, each equally inclined to the three-fold axis. The lattice-constant relations of the trigonal system given in Table 2.2 are based on this primitive rhombohedral cell. However, it has none of the unit cell axes parallel to the characteristic three-fold rotational axis. Therefore, it is more common to take a larger repeat unit, called a *triple hexagonal cell*, as the conventional unit cell of a trigonal crystal (Figure 2.13(c)). This cell has internal lattice points at heights of 1/3 and 2/3 of the repeat distance along the characteristic triad

Geometry of Crystals

axis. The cell is of the same shape as the unit cell of the simple hexagonal Bravais lattice. Therefore, the lattice parameters of a trigonal crystal are often given by $a = b \neq c$, $\alpha = \beta = 90°$, $\gamma = 120°$. It is to be noted that the triple hexagonal cell has nothing to do with the hexagonal system. No matter what the unit cell is chosen, the crystal is trigonal once it exhibits three-fold rotational symmetry only.

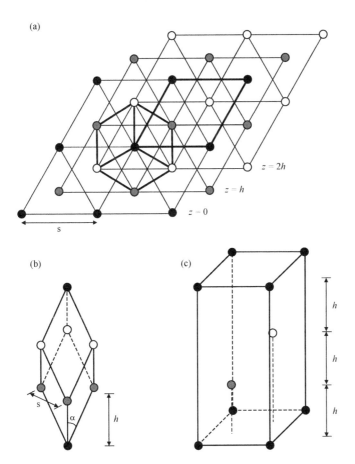

FIGURE 2.13 (a) Diagram showing how a simple trigonal lattice can be constructed by stacking tri-equiangular nets. The first layer is marked with dots, the second layer with gray circles, and the third layer with open circles. The fourth layer overlaps the first layer. s is the separation of lattice points on the layer and h, the inter-layer height. (b) A primitive rhombohedral cell of the trigonal lattice. (c) Triple hexagonal cell.

Three different types of cubic lattices are also built up by stacking tri-equiangular layers. We start with the rhombohedral cell shown in Figure 2.13(b), where s is the separation of lattice points on the layer and h, the height between the layers. In a trigonal crystal, the inter-layer height is unrelated to the separation between lattice points within the layer, which means that three-fold rotational symmetry always exists in the vertical direction regardless of the combination of s and h. If we stack tri-equiangular nets such that $h = s/\sqrt{6}$, the rhombohedron of Figure 2.13(b) becomes a cube with $\alpha = 90°$. This makes the three-fold rotational symmetry to be developed in three other directions. The resulting Bravais lattice is a simple cubic. Similarly, face-centered cubic (FCC) and body-centered cubic (BCC) lattices can be derived by letting $h = 2s/\sqrt{6}$ and $h = s/2\sqrt{6}$, where the values of α are 60° and 109.28°, respectively. The cubic lattices are constructed by stacking tri-equiangular nets along the body diagonal of the conventional unit cell. Figure 2.3 shows the primitive and conventional unit cells of the FCC lattice. If the rhombohedron of Figure 2.13(b) is made to have the same shape as the primitive cell of this FCC lattice, the constructed space lattice possesses four three-fold rotation axes and the crystal system becomes cubic. Figure 2.14 shows the primitive cell of the BCC lattice. It is also a rhombohedron, which is spread over a total of four conventional cells. The height of the rhombohedron is half the body-diagonal length of the conventional cell. Thus, when the cubic cell has a lattice constant a, we obtain the relation of $3h = \sqrt{3}a/2$. Since s is equal to $\sqrt{2}a$, the primitive cell has a geometry of $h = s/2\sqrt{6}$. That the primitive cell of the FCC lattice has $h = 2s/\sqrt{6}$ can be easily found.

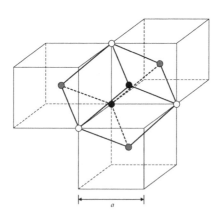

FIGURE 2.14 A primitive cell of the BCC lattice.

Geometry of Crystals

EXAMPLE 2.1.

A certain crystal has the following atomic coordinates within its unit cell ($a = b = c$, $\alpha = \beta = \gamma = 90°$). State the crystal system and Bravais lattice for each case.

(a) **A**: (0,0,0), (1/2,1/2,1/2) **B**: (0,0,1/2), (1/2,1/2, 0)

(b) **A**: (0,0,0), (1/2,1/2,0) **B**: (0,1/2,0)

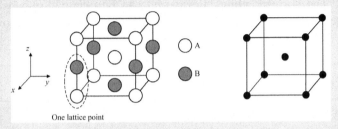

FIGURE 2.15 Atomic arrangements within the unit cell and its lattice in case of (a)

Answer:

Figure 2.15 shows the atomic arrangement within the unit cell in (a). As one **A** atom and one **B** atom forms a single lattice point, the space lattice becomes body-centered. This structure possesses a four-fold rotation axis along the z direction and two-fold axes along the x and y directions. The crystal system is tetragonal and the Bravais lattice is thus body-centered tetragonal. In the case of (b), two **A** atoms and one **B** atom constitute a lattice point, resulting in a simple lattice (Figure 2.16). This structure exhibits three mutually orthogonal two-fold rotation axes. Then the Bravais lattice is simple orthogonal.

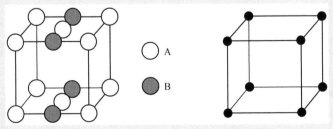

FIGURE 2.16 Atomic arrangements within the unit cell and its lattice in case of (b)

It should be noted that the crystal system and Bravais lattice of a crystal are determined by the symmetry elements that it possesses, not by the unit cell geometry. If a crystal is cubic, its lattice parameters should be $a = b = c$, $\alpha = \beta = \gamma = 90°$. However, the opposite is not always true as shown by this example. Table 2.2 lists the lattice types allowed for each crystal system. The crystal system of a crystal and its Bravais lattice cannot be stated simply by the unit cell dimensions. The corresponding information can be obtained only when the atomic arrangement associated with each lattice point is known.

EXAMPLE 2.2.

Why is there no base-centered tetragonal lattice?

Answer: As shown in Figure 2.17, the base-centered tetragonal lattice is fundamentally the same as the simple lattice. Although we can choose a repeat unit arbitrarily as long as the characteristic four-fold symmetry is maintained, the simple cell is more commonly used than the base-centered cell.

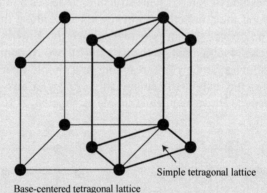

FIGURE 2.17 The relationship between base-centered tetragonal lattice and simple tetragonal lattice.

Geometry of Crystals

2.4 DIRECTIONS AND PLANES

2.4.1 INDICES OF DIRECTION

A direction in the crystal is specified as follows. When a vector **r** is drawn from one corner of the unit cell, it can be represented as $\mathbf{r} = u\mathbf{a} + v\mathbf{b} + w\mathbf{c}$, where **a**, **b**, and **c** are the unit cell vectors having lengths equal to the lattice parameters (Figure 2.18(a)). This vector has components $u\mathbf{a}$, $v\mathbf{b}$, and $w\mathbf{c}$ along the x, y, and z axes of the crystal, respectively, and its direction is then denoted as $[uvw]$. No matter what the values of u, v, and w, the triple indices representing a direction are given by a set of smallest integers. Although $[\frac{1}{2}\frac{1}{2}1]$, [112], and [336] all represent the same direction, [112] is the preferred form. The indices are enclosed in square brackets and negative indices are written with a bar over the number. Some direction notations are shown in Figure 2.18(b). Directions related by symmetry are called *directions of a form*. The directions that belong to a form can be represented by a single set of indices enclosed in angular brackets. In a cubic crystal, the four body-diagonal directions given by [111], [$\bar{1}$11], [1$\bar{1}$1], and [$\bar{1}\bar{1}$1] are crystallographically identical. Thus, these directions may be represented by the form <111>. Likewise, [101] and [110] belong to the same form symbolized by <110>. However, [101] is not equivalent to [110] in the tetragonal system.

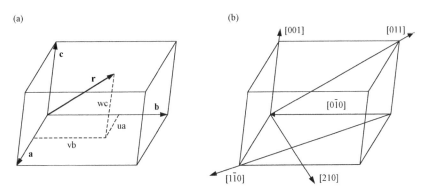

FIGURE 2.18 (a) The direction of a vector given by $\mathbf{r} = v\mathbf{a}+v\mathbf{b}+w\mathbf{c}$ is denoted as $[uvw]$. The triple indices representing a direction are always cleared of fractions and given by a set of smallest integers. (b) Indices of directions

2.4.2 INDICES OF PLANE

Crystallographic planes can also be indexed using a system developed by W. H. Miller. When a plane makes intercepts at a/h, b/k, and c/l with the crystallographic axes, it is represented by (hkl) where the Miller indices h, k, and l are enclosed in parentheses (Figure 2.19(a)). The negative index is marked with a bar over it. If the given plane is parallel to a crystallographic axis, its intercept on that axis is regarded as infinity and the corresponding Miller index is zero. Although Figure 2.19(a) depicts a single plane nearest the origin, the Miller indices (hkl) refer to the whole set of parallel equidistant planes as shown in Figure 2.19(b). The whole set of (hkl) planes is thus given by

$$\frac{hx}{a} + \frac{ky}{b} + \frac{lz}{c} = m \tag{2.1}$$

where m is an integer. If $m = 0$, the (hkl) plane passes through the origin; if $m = 1$, the plane makes intercepts at a/h, b/k, and c/l with the crystallographic axes; if $m = 2$, the intercepts are $2a/h$, $2b/k$, and $2c/l$; and if $m = -1$, the intercepts are $-a/h$, $-b/k$, and $-c/l$. The ($nh\ nk\ nl$) planes are parallel to the (hkl) planes and have $1/n$th the spacing.

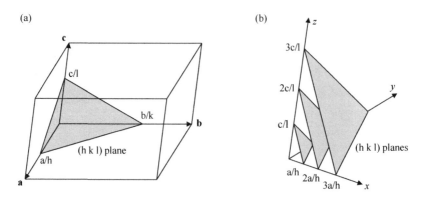

FIGURE 2.19 (a) Plane designation by Miller indices h, k, and l. (b) Whole set of (hkl) planes.

Geometry of Crystals

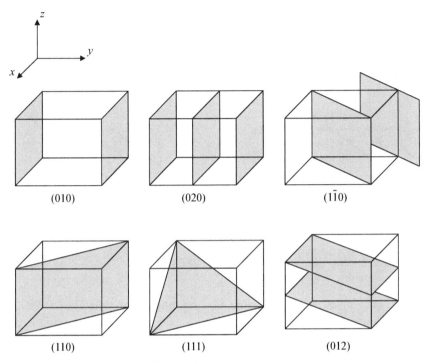

FIGURE 2.20 Miller indices of some planes.

Some examples of the Miller index notation are given in Figure 2.20. It is important to note that negative Miller indices are not associated with the negative intercepts. Figure 2.21(a) shows a plane extending normal to the page. For simplicity, we assume that this plane is parallel to the z axis. Since the lattice is periodically arrayed, any corner of the unit cell can be taken as the origin. If we consider the point O_1 of Figure 2.21(a) as an origin, the given plane will be denoted by (110). However, its notation changes to ($\bar{1}\bar{1}0$) if we consider the point O_2 as a new origin. Once the positive directions of the crystallographic axes are determined, the indexing of a fixed plane should be invariant no matter which lattice point is chosen as the origin. Any plane has inner and outer sides. The Miller indexing system is in such a way that h, k, and l indices derived from the intercept values are used to represent the outer side of the given plane.

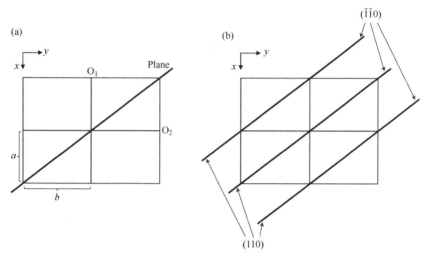

FIGURE 2.21 Definition of (hkl) and $(\bar{h}\bar{k}\bar{l})$ planes.

When viewed from the origin O_1, the plane shown in Figure 2.21(a) have intercepts of a and b on the x and y axes. Thus, its outer side is denoted by (110) and the inner side, $(\bar{1}\bar{1}0)$. If the point O_2 is set as a new origin, the intercepts are $-a$ and $-b$. Likewise, the obtained Miller indices $(\bar{1}\bar{1}0)$ are dedicated to the outer side of the plane, which was formerly the inner side. Therefore, one side of the whole set of planes is represented by (110) and the other, $(\bar{1}\bar{1}0)$, as shown in Figure 2.21(b). If the front surface of a single crystal wafer has indices (hkl), its rear surface will be $(\bar{h}\bar{k}\bar{l})$. Two surfaces may or may not exhibit the same properties depending on the crystal symmetry. While the surface properties of (111) and $(\bar{1}\bar{1}\bar{1})$ are identical in Si, they are different in GaAs due to the absence of a center of symmetry. Many physical and chemical properties of a crystal are influenced by its symmetry. The inversion symmetry, characterized by the presence of a center of symmetry, makes the properties of a crystal along a certain direction and its opposite direction equalized. Unlike Si, GaAs lacks a center of symmetry. Therefore, GaAs has different etching rates on the (111) and $(\bar{1}\bar{1}\bar{1})$ surfaces whose normals are oppositely directed. The presence of a center of symmetry is beneficial for some applications but not for others. A case is discussed in Example 2.3. In crystals, there are sets of equivalent planes related by symmetry. They are called *planes of a form*, and the indices of any one plane, enclosed in braces, stands for the whole set of planes. For example, (211), ($\bar{1}$21), ($\bar{2}\bar{1}$1), and (1$\bar{2}$1) in a tetragonal crystal are planes of the form {211}. As shown in Figure 2.22, all of them can be generated from any one by operation of the four-fold rotation axis.

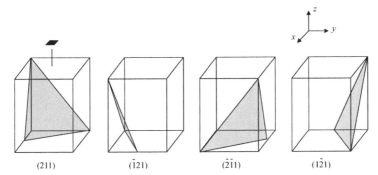

FIGURE 2.22   Planes of the form {211} in a tetragonal crystal.

The unit cell of the hexagonal and trigonal systems is defined by two equal and coplanar vectors $\mathbf{a}_1$ and $\mathbf{a}_2$ at 120°, and $\mathbf{c}$ at right angles (Figure 2.23). These two crystal systems differ from the other systems in that the operation of their characteristic symmetry generates a third axis $\mathbf{a}_3$ equivalent to $\mathbf{a}_1$ and $\mathbf{a}_2$ axes. To take into account this extra axis, a fourth index i is often introduced into the plane indexing system. Thus, a plane in the hexagonal and trigonal systems is represented by (*hkil*), which are known as *Miller-Bravais indices*. Since the intercept of a plane on $\mathbf{a}_3$ is determined by the intercepts on $\mathbf{a}_1$ and $\mathbf{a}_2$, the value of i depends on h and k, with the relation of $h + k = -i$. Figure 2.23 shows some planes denoted by the Miller-Bravais indices. The reason for introducing the Miller-Bravais indices is to give similar indices to similar planes. It can be simply demonstrated by considering the form {10$\bar{1}$0} in a hexagonal crystal. {10$\bar{1}$0} includes (10$\bar{1}$0), (01$\bar{1}$0), ($\bar{1}$100), ($\bar{1}$010), (0$\bar{1}$10), and (1$\bar{1}$00). These represent the side faces of a hexagonal prism, which are symmetrically located with respect to the rotation axis $\mathbf{c}$. The Miller-Bravais indices contain the same quartet of numbers regularly interchanged in position and sign. Thus, the symmetry relationship between the faces is immediately obvious. In contrast, the Miller indices do not immediately imply this relationship, since the corresponding planes are denoted by (100), (010), ($\bar{1}$10), ($\bar{1}$00), (0$\bar{1}$0), and (1$\bar{1}$0). Nevertheless, many people still feel more familiar with the Miller indices. Whether using the Miller indices or the Miller-Bravais indices is indeed a matter of choice.

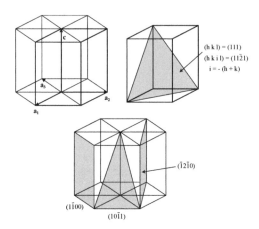

FIGURE 2.23 Plane designation by Miller-Bravais indices (*hkil*) in the hexagonal system.

EXAMPLE 2.3. Center of symmetry vs. electro-optic effect

For some applications, the crystal should not possess a center of symmetry. An example is given here. The electro-optic effect refers to the phenomenon that the refractive index of a material changes in response to an electric field. This effect is widely used in many electronic and optical devices. A function *f(x)* with variable *x* can be expanded into the following power series.

$$f(x) = f(0) + f'(0)x + \frac{f''(0)}{2}x^2 + \ldots \quad (2.2)$$

The refractive index of a medium is a function *n(E)* of the applied electric field *E*. Since this function varies slightly with *E*, it can be expanded in a power series about *E* = 0.

$$n(E) = n(0) + a_1 E + \frac{a_2}{2}E^2 + \ldots \quad (2.3)$$

Geometry of Crystals

where a_1 and a_2 represent the first-order and second-order electro-optic coefficients. In usual, the second-order coefficient is much smaller than the first-order term. When an electric field with magnitude E is applied in the positive z direction of a crystal, the refractive index is given by Eq. (2.3). If an electric field of the same magnitude is applied in the negative z direction, it will be expressed by

$$n(-E) = n(0) - a_1 E + \frac{a_2}{2} E^2 + ... \qquad (2.4)$$

If this crystal possesses a center of symmetry, the refractive index change should be the same in both cases, i.e., $n(E) = n(-E)$. This leads to $a_1 = 0$. Thereby, the crystal contains only the second-order coefficient. To be an effective electro-optic medium, a center of symmetry should be absent from the crystal.

2.5 RECIPROCAL LATTICE

2.5.1 INTRODUCTION

The reciprocal lattice literally means the reciprocal-space version of a real lattice. The reciprocal lattice of a reciprocal lattice is the original lattice. The concept of this reciprocal lattice was first introduced by the German physicist and crystallographer H. Ewald in 1921. The reciprocal lattice plays an essential role in most analytic studies of periodic structures, particularly in the theory of diffraction. Although the reciprocal lattice may appear abstract at first glance, it is very useful to calculate the angle between planes in a crystal. As will be discussed later, the direction of X-ray diffraction is governed by the Bragg law. However, there are some effects such as off-Bragg angle diffraction that cannot be explained by the simple Bragg law. A more general theory of diffraction is required for the explanation of these effects. The reciprocal lattice theory has become indispensable in describing the behaviors of electrons in crystals. Moreover, the reciprocal lattice of a crystal can be directly visualized by electron diffraction, implying that it is no longer an abstract concept. Each point hkl in the reciprocal lattice corresponds to a set of lattice planes (hkl) in the real lattice. The direction of the reciprocal lattice vector is normal to the real space planes. The magnitude of the reciprocal lattice vector is given in reciprocal length and is equal to the reciprocal of the interplanar spacing of the real planes. The simple cubic Bravais lattice, with side a, has a simple cubic reciprocal lattice of side $1/a$. This is based on the crystallographer's definition. In the "solid state physics" or "optics" definition, the corresponding reciprocal lattice has a side of $2\pi/a$. Even in the latter case, the factor of 2π is common in all expressions and does not alter the physical implication of the reciprocal lattice concept. The reciprocal lattice to an FCC lattice becomes a BCC lattice and the reciprocal lattice to a BCC lattice, an FCC lattice. The Bragg condition on X-ray diffraction in real space predicts only the possible directions of diffraction and mentions nothing about the diffraction intensity. Once the reciprocal lattice of a crystal structure is constructed, not only the diffraction direction but also its intensity can be predicted from the Bragg condition in reciprocal space. Since the reciprocal lattice is formulated with the products of vectors, we take a brief look at the fundamental vector analysis.

2.5.2 VECTOR ANALYSIS

A vector is an entity possessing both magnitude and direction. If a vector **A** has components A_x, A_y, A_z along the x, y, z coordinate axes (Figure 2.24), it is given by

$$\mathbf{A} = A_x\mathbf{i} + A_y\mathbf{j} + A_z\mathbf{k} \tag{2.5}$$

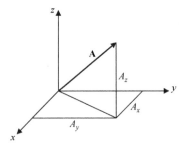

FIGURE 2.24 A vector given by $\mathbf{A} = A_x\mathbf{i} + A_y\mathbf{j} + A_z\mathbf{k}$, where **i**, **j**, and **k** are unit vectors in the directions of x, y, and z axes.

Geometry of Crystals

where **i**, **j**, **k** are unit vectors in the directions of x, y, z axes. Another vector **B** can be represented in the same way with its components. The scalar product of two vectors **A** and **B** (also called the dot product) is expressed as

$$\mathbf{A} \cdot \mathbf{B} = AB \cos \theta \tag{2.6}$$

where A and B are the magnitudes of **A** and **B** vectors, respectively, and θ is the angle between them. It follows from Eq. (2.6) that the scalar product is commutative, i.e., $\mathbf{A} \cdot \mathbf{B} = \mathbf{B} \cdot \mathbf{A}$. The scalar product $\mathbf{A} \cdot \mathbf{B}$ is equal to the projection of **A** on the direction of **B** multiplied by the magnitude of **B**. The scalar product of a vector with itself is equal to the square of its magnitude. When the coordinate axes are orthogonal to one another, this gives

$$|\mathbf{A}| = \sqrt{A_x^2 + A_y^2 + A_z^2} \tag{2.7}$$

For orthogonal axes, the unit vectors **i**, **j**, **k** have the following relations.

$$\begin{aligned} \mathbf{i} \cdot \mathbf{i} = \mathbf{j} \cdot \mathbf{j} = \mathbf{k} \cdot \mathbf{k} = 1 \\ \mathbf{i} \cdot \mathbf{j} = \mathbf{j} \cdot \mathbf{k} = \mathbf{k} \cdot \mathbf{i} = 0 \end{aligned} \tag{2.8}$$

Thus, the above Eq. (2.6) becomes

$$\mathbf{A} \cdot \mathbf{B} = \left(A_x \mathbf{i} + A_y \mathbf{j} + A_z \mathbf{k} \right) \cdot \left(B_x \mathbf{i} + B_y \mathbf{j} + B_z \mathbf{k} \right) = A_x B_x + A_y B_y + A_z B_z \tag{2.9}$$

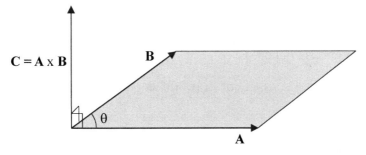

FIGURE 2.25 The vector product **A**×**B** is a vector perpendicular to the plane containing both **A** and **B**. The magnitude of **A**×**B** is AB $sin\theta$.

The vector product of two vectors **A** and **B**, denoted by **A**×**B**, is a vector perpendicular to the plane containing both **A** and **B**. Let **C** be the vector product of these two vectors, i.e., **C** = **A**×**B**. The direction of **C** is such that **A, B, C**, in this order, forms a right-handed screw as shown in Figure 2.25. The direction of the vector product is reversed if the order of multiplication is reversed, that is, **B**×**A** = −(**A**×**B**). The magnitude of **A**×**B** is $AB \sin\theta$, which is equal to the area of the parallelogram with **A** and **B** as adjacent sides. If two vectors are parallel, their vector product vanishes. If they are normal to each other, the magnitude of the vector product is equal to the product of their magnitudes. Thus, when the coordinate axes are rectangular,

$$\mathbf{i}\times\mathbf{i} = \mathbf{j}\times\mathbf{j} = \mathbf{k}\times\mathbf{k} = 0$$
$$\mathbf{i}\times\mathbf{j} = \mathbf{k}, \mathbf{j}\times\mathbf{k} = \mathbf{i}, \mathbf{k}\times\mathbf{i} = \mathbf{j} \tag{2.10}$$

Using these relations, we obtain the vector product of **A** and **B** vectors as given below.

$$\mathbf{A}\times\mathbf{B} = \left(A_x\mathbf{i} + A_y\mathbf{j} + A_z\mathbf{k}\right)\times\left(B_x\mathbf{i} + B_y\mathbf{j} + B_z\mathbf{k}\right) = (A_yB_z - A_zB_y)\mathbf{i} + (A_zB_x - A_xB_z)\mathbf{j} + (A_xB_y - A_yB_x)\mathbf{k} \tag{2.11}$$

The relationship of Eq. (2.11) may be more easily memorized by expressing it in the form of a matrix.

$$\mathbf{A}\times\mathbf{B} = \begin{vmatrix} \mathbf{i} & \mathbf{j} & \mathbf{k} \\ A_x & A_y & A_z \\ B_x & B_y & B_z \end{vmatrix} \tag{2.12}$$

The product $\mathbf{C}\cdot(\mathbf{A}\times\mathbf{B})$ between three nonparallel vectors is a scalar whose magnitude represents the volume of a parallelepiped formed by these vectors. Since the volume of the parallelepiped is fixed, the following relations are obtained.

$$\mathbf{C}\cdot(\mathbf{A}\times\mathbf{B}) = \mathbf{A}\cdot(\mathbf{B}\times\mathbf{C}) = \mathbf{B}\cdot(\mathbf{C}\times\mathbf{A}) \tag{2.13}$$

2.5.3 *RECIPROCAL LATTICE*

Suppose that the real lattice has a unit cell defined by the vectors **a**, **b**, and **c**, as shown in Figure 2.26. Then, the reciprocal lattice has a unit cell defined by the vectors **a***, **b***, and **c***, where

Geometry of Crystals

$$\mathbf{a}^* = \frac{\mathbf{b} \times \mathbf{c}}{V}, \mathbf{b}^* = \frac{\mathbf{c} \times \mathbf{a}}{V}, \mathbf{c}^* = \frac{\mathbf{a} \times \mathbf{b}}{V} \tag{2.14}$$

and V is the volume of unit cell of the real lattice. From these relations, we note that $\mathbf{c}^*$ is a vector normal to the plane containing both $\mathbf{a}$ and $\mathbf{b}$. With reference to Figure 2.26, the vector product $\mathbf{a} \times \mathbf{b}$ has a magnitude equal to the area of basal plane of the given unit cell. Thus, the magnitude of $\mathbf{c}^*$ is equivalent to the reciprocal of the cell height. Its magnitude is simply the reciprocal of the spacing of the (001) planes in the real lattice, i.e., $|\mathbf{c}^*| = 1/d_{001}$. Similarly, we find that $\mathbf{a}^*$ and $\mathbf{b}^*$ are normal to the (100) and (010) planes, respectively, of the real lattice. Once $\mathbf{a}^*$, $\mathbf{b}^*$, and $\mathbf{c}^*$ are obtained, the whole reciprocal lattice can be constructed by repeated translation of the unit cell defined by these vectors.

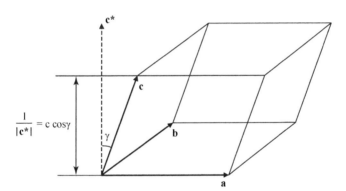

FIGURE 2.26 Definition of the reciprocal lattice vector $\mathbf{c}^*$.

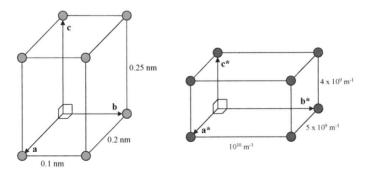

FIGURE 2.27 Relation of a real lattice to its reciprocal lattice.

56   X-Ray Diffraction for Materials Research: From Fundamentals to Applications

As an example, Figure 2.27 shows a real lattice and its reciprocal lattice. From Eq. (2.13) and (2.14), we note that

$$\mathbf{a} \cdot \mathbf{a}^* = \mathbf{b} \cdot \mathbf{b}^* = \mathbf{c} \cdot \mathbf{c}^* = 1$$
$$\mathbf{a} \cdot \mathbf{b}^* = \mathbf{b} \cdot \mathbf{c}^* = \mathbf{c} \cdot \mathbf{a}^* = 0 \qquad (2.15)$$

A direction in the reciprocal lattice can also be represented with a vector drawn from the origin to any lattice point having indices hkl. This vector $\mathbf{H}_{hkl}$ is expressed with its coordinates in terms of the basic unit cell vectors.

$$\mathbf{H}_{hkl} = h\mathbf{a}^* + k\mathbf{b}^* + l\mathbf{c}^* \qquad (2.16)$$

where h, k, and l are integers. The reciprocal lattice points are labeled with these indices. For instance, the point at the end of the $\mathbf{a}^*$ vector is labeled 100, and that at the end of $\mathbf{a}^* + 2\mathbf{b}^*$, 120. While h, k, and l are plane indices in the real lattice, they are here used to denote a direction. This is because the plane in a real lattice is interrelated with the direction in its reciprocal lattice as follows.

1. The $\mathbf{H}_{hkl}$ vector is perpendicular to the (hkl) plane of the real lattice.
2. The length of $\mathbf{H}_{hkl}$ is equal to the reciprocal of the spacing of the (hkl) planes, i.e., $H_{hkl} = 1/d_{hkl}$.

It follows from these relations that each reciprocal lattice point is related to a set of lattice planes in the crystal and represents the orientation and spacing of such a set of planes. Some examples of the reciprocal lattice are provided before proving the above statements. Let's consider the cubic unit cell of a cubic crystal and its reciprocal lattice as shown in Figure 2.28(a). For any crystal whose unit cell is defined by three mutually perpendicular vectors, i.e., cubic, tetragonal, and orthorhombic, the basic vectors $\mathbf{a}^*$, $\mathbf{b}^*$, and $\mathbf{c}^*$ of the reciprocal lattice are parallel to $\mathbf{a}$, $\mathbf{b}$, and $\mathbf{c}$, respectively, and their magnitudes are simply the reciprocals of a, b, and c. Figure 2.28(a) shows the case for cubic crystal, where $|\mathbf{a}| = |\mathbf{b}| = |\mathbf{c}| = a$, $|\mathbf{a}^*| = |\mathbf{b}^*| = |\mathbf{c}^*| = 1/a$. Some planes with low Miller indices are shown in Figure 2.28(a), together with the corresponding $\mathbf{H}$ vectors. It looks obvious that $\mathbf{H}_{100}$, $\mathbf{H}_{010}$, and $\mathbf{H}_{210}$ are perpendicular to (100), (010), and (210) planes, respectively. Since the (220) planes have half the spacing of the (110) planes, $\mathbf{H}_{220}$ is twice as long as $\mathbf{H}_{110}$. In the hexagonal crystal, two equal-magnitude $\mathbf{a}$ and $\mathbf{b}$ vectors makes an

Geometry of Crystals

angle of 120°, both of which are at right angles with another vector **c**. Thus, the angle between **a*** and **b*** becomes 60° and **c*** remains parallel to **c**, as depicted in Figure 2.28(b). $\mathbf{H}_{110}$ and $\mathbf{H}_{120}$ are also perpendicular to (110) and (120) planes, respectively, in this nonrectangular coordinate system.

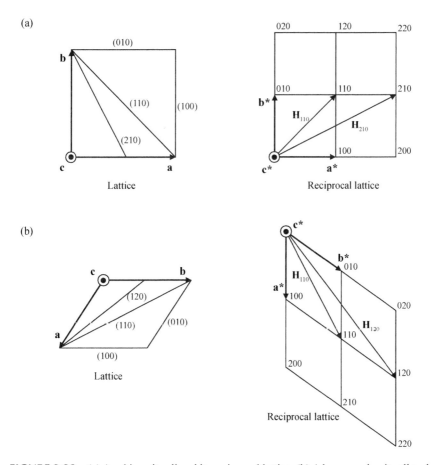

FIGURE 2.28 (a) A cubic unit cell and its reciprocal lattice. (b) A hexagonal unit cell and its reciprocal lattice. The reciprocal lattice vector $\mathbf{H}_{hkl}$ is always perpendicular to the (*hkl*) plane of the real lattice.

The fact that $\mathbf{H}_{hkl}$ is normal to (hkl) and $\mathbf{H}_{hkl}$ is the reciprocal of d_{hkl} is valid for all crystal systems. Now we prove them with Figure 2.29, where an (hkl) plane nearest the origin makes intercepts at A, B, and C with the unit cell vectors $\mathbf{a}$, $\mathbf{b}$, and $\mathbf{c}$. From the definitions of Miller indices, the vectors from the origin to the point A, B, and C are $\mathbf{a}/h$, $\mathbf{b}/k$, and $\mathbf{c}/l$, respectively. Then, the vector drawn from A to B is given by $\mathbf{AB} = \mathbf{b}/k - \mathbf{a}/h$. Similarly, we have $\mathbf{AC} = \mathbf{c}/l - \mathbf{a}/h$. If $\mathbf{H}_{hkl}$ is normal to (hkl), it must be normal to both $\mathbf{AB}$ and $\mathbf{AC}$, which are nonparallel vectors lying in the same plane. Carrying out the dot product of $\mathbf{H}_{hkl}$ with these two vectors, we find

$$\mathbf{H}_{hkl} \cdot \mathbf{AB} = \left(h\mathbf{a}^* + k\mathbf{b}^* + l\mathbf{c}^*\right) \cdot \left(\mathbf{b}/k - \mathbf{a}/h\right) = -\mathbf{a}^* \cdot \mathbf{a} + \mathbf{b}^* \cdot \mathbf{b} = 0$$

$$\mathbf{H}_{hkl} \cdot \mathbf{AC} = \left(h\mathbf{a}^* + k\mathbf{b}^* + l\mathbf{c}^*\right) \cdot \left(\mathbf{c}/l - \mathbf{a}/h\right) = -\mathbf{a}^* \cdot \mathbf{a} + \mathbf{c}^* \cdot \mathbf{c} = 0 \quad (2.17)$$

Since the products are zero, $\mathbf{H}_{hkl}$ are normal to $\mathbf{AB}$ and $\mathbf{AC}$, and also to the (hkl) plane containing these two vectors. Let's assume that the reciprocal lattice vector $\mathbf{H}_{hkl}$ meets with the (hkl) plane at point N. As $\mathbf{H}_{hkl}$ is perpendicular to (hkl), the distance from the origin to the point N, $\overline{ON}$, is equal to the spacing of the (hkl) planes, d_{hkl}. This value can be obtained by projecting $\mathbf{OA} = \mathbf{a}/h$ onto the direction of $\mathbf{H}_{hkl}$. $\mathbf{H}_{hkl}/H_{hkl}$ is a unit vector in the direction of $\mathbf{H}_{hkl}$. Therefore, d_{hkl} is given by

$$d_{hkl} = \overline{ON} = \mathbf{OA} \cdot \frac{\mathbf{H}_{hkl}}{H_{hkl}} = \frac{\mathbf{a}}{h} \cdot \frac{\left(h\mathbf{a}^* + k\mathbf{b}^* + l\mathbf{c}^*\right)}{H_{hkl}} = \frac{1}{H_{hkl}} \quad (2.18)$$

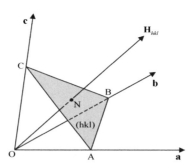

FIGURE 2.29 Relation between the reciprocal lattice vector $\mathbf{H}_{hkl}$ and the lattice plane (hkl).

Geometry of Crystals

The reciprocal lattice is very useful for solving many problems in crystal geometry. *Planes of a zone* are planes that are all parallel to a line, called the *zone axis* (Figure 2.30). For instance, the planes (100), (110), and (120) are all parallel to the direction [001]. They would be said to lie in the zone [001]. Such planes may have different indices and spacings. When the zone axis has indices [uvw], any planes belong to that zone if their indices (hkl) satisfy the relation

$$hu + kv + lw = 0 \tag{2.19}$$

This is the condition that the normal to (hkl) is perpendicular to the direction [uvw]; the dot product of $\mathbf{r} = u\mathbf{a} + v\mathbf{b} + w\mathbf{c}$ and $\mathbf{H}_{hkl} = h\mathbf{a}^* + k\mathbf{b}^* + l\mathbf{c}^*$ should be zero. There is a zone axis for any two nonparallel planes because they are both parallel to the line of intersection. If their indices are $(h_1 k_1 l_1)$ and $(h_2 k_2 l_2)$, the zone axis [uvw] is given by

$$u = k_1 l_2 - k_2 l_1$$
$$v = l_1 h_2 - l_2 h_1$$
$$w = h_1 k_2 - h_2 k_1 \tag{2.20}$$

These relations can be obtained by solving two (2.19) equations simultaneously for u, v, and w. Since the zone axis represents a direction, we are concerned with the relative ratios between u, v, and w, not the absolute values.

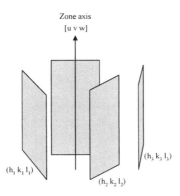

FIGURE 2.30 Zone axis.

60 X-Ray Diffraction for Materials Research: From Fundamentals to Applications

The reciprocal lattice is also very convenient in calculating the planar spacing and interplanar angles. It has been shown in Eq. (2.18) that the perpendicular distance between successive (hkl) planes, d_{hkl}, is equal to the inverse of the magnitude of the corresponding reciprocal lattice vector;

$$d_{hkl} = \frac{1}{\left| h\mathbf{a}^* + k\mathbf{b}^* + l\mathbf{c}^* \right|} \tag{2.21}$$

The interplanar spacing d_{hkl} can simply be calculated by forming the scalar product of the reciprocal lattice vector with itself.

$$
\begin{aligned}
\left| h\mathbf{a}^* + k\mathbf{b}^* + l\mathbf{c}^* \right|^2 &= \left(h\mathbf{a}^* + k\mathbf{b}^* + l\mathbf{c}^* \right) \cdot \left(h\mathbf{a}^* + k\mathbf{b}^* + l\mathbf{c}^* \right) \\
&= h^2 \mathbf{a}^* \cdot \mathbf{a}^* + k^2 \mathbf{b}^* \cdot \mathbf{b}^* + l^2 \mathbf{c}^* \cdot \mathbf{c}^* + 2kl\mathbf{b}^* \cdot \mathbf{c}^* + 2lh\mathbf{c}^* \cdot \mathbf{a}^* + 2hk\mathbf{a}^* \cdot \mathbf{b}^*
\end{aligned}
\tag{2.22}
$$

These expressions are very easy to use when the unit cell axes are orthogonal as in the cubic, tetragonal, and orthorhombic systems: $\left| \mathbf{a}^* \right| = 1/a, \left| \mathbf{b}^* \right| = 1/b, \left| \mathbf{c}^* \right| = 1/c$, and $\mathbf{b}^* \cdot \mathbf{c}^* = \mathbf{c}^* \cdot \mathbf{a}^* = \mathbf{a}^* \cdot \mathbf{b}^* = 0$. In the cubic system, $a = b = c$, while the tetragonal system has $a = b \neq c$. The lattice parameters a, b, and c are in general different from one another in the orthorhombic system. For the hexagonal system, $\left| \mathbf{a}^* \right| = \left| \mathbf{b}^* \right| = 2/\sqrt{3}a$, $\left| \mathbf{c}^* \right| = 1/c$, $\mathbf{b}^* \cdot \mathbf{c}^* = \mathbf{c}^* \cdot \mathbf{a}^* = 0$, and $\mathbf{a}^* \cdot \mathbf{b}^* = 2/3a^2$. Thus, the value of d_{hkl} is given by the following equations.

$$(\text{Cubic})\, d_{hkl} = \frac{a}{\sqrt{h^2 + k^2 + l^2}}$$

$$(\text{Tetragonal})\, d_{hkl} = \frac{a}{\sqrt{h^2 + k^2 + (\frac{a^2}{c^2})l^2}}$$

$$(\text{Orthorhombic})\, d_{hkl} = \frac{a}{\sqrt{h^2 + (\frac{a^2}{b^2})k^2 + (\frac{a^2}{c^2})l^2}}$$

$$(\text{Hexagonal})\, d_{hkl} = \frac{a}{\sqrt{4(h^2 + hk + k^2)/3a^2 + l^2/c^2}} \tag{2.23}$$

Geometry of Crystals

61

It is to be noted that the interplanar spacing is not affected by the lattice type. For instance, simple cubic lattice and FCC lattice have the same d_{hkl} value if their lattice constant a is the same. The trigonal system possesses three-fold rotational symmetry only. Nevertheless, a triple hexagonal cell rather than its primitive rhombohedral cell is more commonly employed as the conventional unit cell. When the plane indices hkl are based on this triple hexagonal cell, the interplanar spacing is expressed by the same way as in the hexagonal system. The angle between two nonparallel planes is identical to the angle between their normals, i.e., the angle between the corresponding reciprocal lattice vectors. Then, the interplanar angle θ between $(h_1 k_1 l_1)$ and $(h_2 k_2 l_2)$ is given by

$$\cos\theta = \frac{\left(h_1\mathbf{a}^* + k_1\mathbf{b}^* + l_1\mathbf{c}^*\right)\cdot\left(h_2\mathbf{a}^* + k_2\mathbf{b}^* + l_2\mathbf{c}^*\right)}{\left|h_1\mathbf{a}^* + k_1\mathbf{b}^* + l_1\mathbf{c}^*\right|\left|h_2\mathbf{a}^* + k_2\mathbf{b}^* + l_2\mathbf{c}^*\right|} \tag{2.24}$$

This general formula can also be written out for each crystal system.

$$(\text{Cubic})\cos\theta = \frac{h_1 h_2 + k_1 k_2 + l_1 l_2}{\sqrt{h_1^2 + k_1^2 + l_1^2}\sqrt{h_2^2 + k_2^2 + l_2^2}}$$

$$(\text{Tetragonal})\cos\theta = \frac{h_1 h_2 + k_1 k_2 + (a^2/c^2) l_1 l_2}{\sqrt{h_1^2 + k_1^2 + (a^2/c^2) l_1^2}\sqrt{h_2^2 + k_2^2 + (a^2/c^2) l_2^2}}$$

$$(\text{Orthorhombic})\cos\theta = \frac{\dfrac{h_1 h_2}{a^2} + \dfrac{k_1 k_2}{b^2} + \dfrac{l_1 l_2}{c^2}}{\sqrt{\dfrac{h_1^2}{a^2} + \dfrac{k_1^2}{b^2} + \dfrac{l_1^2}{c^2}}\sqrt{\dfrac{h_2^2}{a^2} + \dfrac{k_2^2}{b^2} + \dfrac{l_2^2}{c^2}}}$$

$$(\text{Hexagonal})\cos\theta = \frac{h_1 h_2 + k_1 k_2 + \dfrac{1}{2}(h_1 k_2 + k_1 h_2) + (3a^2/4c^2) l_1 l_2}{\sqrt{h_1^2 + k_1^2 + h_1 k_1 + (3a^2/4c^2) l_1^2}\sqrt{h_2^2 + k_2^2 + h_2 k_2 + (3a^2/4c^2) l_2^2}} \tag{2.25}$$

EXAMPLE 2.4.

In a tetragonal crystal, the angle between (011) and (123) is 15°. Calculate the ratio of the axis lengths, c/a, and the angle between [123] and [102].

Answer:

The angle between (011) and (123) is equal to that between the corresponding reciprocal lattice vectors $\mathbf{H}_{011} = \mathbf{b}^* + \mathbf{c}^*$ and $\mathbf{H}_{123} = \mathbf{a}^* + 2\mathbf{b}^* + 3\mathbf{c}^*$. The scalar product of these two vectors leads to the following relation.

$$\left(\mathbf{b}^* + \mathbf{c}^*\right) \cdot \left(\mathbf{a}^* + 2\mathbf{b}^* + 3\mathbf{c}^*\right) = \frac{2}{a^2} + \frac{3}{c^2} = \sqrt{\frac{1}{a^2} + \frac{1}{c^2}}\sqrt{\frac{1}{a^2} + \frac{4}{a^2} + \frac{9}{c^2}} \times \cos 15°$$

By multiplying both sides by a^2 and solving the equation, we obtain $c/a = 0.667$. The angle θ between [123] and [102] can be calculated from the dot product of two vectors a + 2b + 3c and a + 2c.

$$a^2 + 6c^2 = \sqrt{5a^2 + 9c^2}\sqrt{a^2 + 4c^2} \times \cos\theta$$

where $\theta = 42.82°$.

EXAMPLE 2.5.

Two different single crystal slabs are stacked together as shown in Figure 2.31. Crystal **A** is tetragonal with $a = 4.0$ Å and $c = 5.2$ Å, and **B** is cubic with $a = 4.0$ Å.

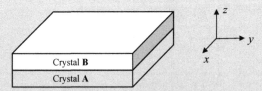

FIGURE 2.31 Two different single crystal slabs stacked together.

(a) What is the angle between $[124]_A$ and $[111]_B$?
(b) Calculate the angle between $(211)_A$ and $(111)_B$
(c) Repeat (a) and (b) when only **B** is counterclockwise rotated by 45° around the z axis.

Geometry of Crystals

Answer:

(a) Since the unit cell axes, x, y, and z, are the same in both crystals, the angle between two directions can be calculated from the dot product of the corresponding vectors. $[124]_A$ is given by the vector $\mathbf{r}_A = \mathbf{a}_A + 2\mathbf{b}_A + 4\mathbf{c}_A$ and $[111]_B$, by $\mathbf{r}_B = \mathbf{a}_B + \mathbf{b}_B + \mathbf{c}_B$. Then,

$$\mathbf{r}_A \cdot \mathbf{r}_B = a_A a_B + 2a_A a_B + 4c_A a_B = \sqrt{5a_A^2 + 16c_A^2}\sqrt{3a_B^2} \times \cos\theta$$

where a_A, c_A, and a_B are the lattice parameters of **A** and **B** crystals. By inserting $a_A = 4\text{Å}$, $c_A = 5.2\text{Å}$, and $a_B = 4\text{Å}$ into the above equation, we obtain $\theta = 33.2°$.

(b) The angle between two planes can be obtained from the dot product of the corresponding reciprocal lattice vectors $\mathbf{H}_A = 2\mathbf{a}_A^* + \mathbf{b}_A^* + \mathbf{c}_A^*$ and $\mathbf{H}_B = \mathbf{a}_B^* + \mathbf{b}_B^* + \mathbf{c}_B^*$.

$$\mathbf{H}_A \cdot \mathbf{H}_B = \frac{2}{a_A a_B} + \frac{1}{a_A a_B} + \frac{1}{c_A a_B} = \sqrt{\frac{5}{a_A^2} + \frac{1}{c_A^2}}\sqrt{\frac{3}{a_B^2}} \times \cos\theta$$

From this relation, $\theta = 23.07°$.

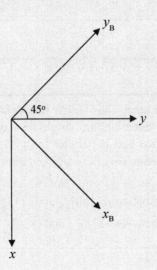

FIGURE 2.32 Unit cell axes of the crystal B before and after rotation.

(c) Since the crystal **B** is rotated, their axes are also rotated. Suppose that x_B and y_B denote the unit cell axes of the crystal **B**. These axes are counterclockwise rotated by $45°$ from the x and y directions, respectively, which represent the unit cell axes of the crystal **A** (Figure 2.32). $[111]_B$ is now based on the new axes, x_B and y_B. We need to express $[111]_B$ with respect to the x and y coordinates in order to calculate the angle between two directions. $[111]_B$ has no component along x and its component along y is $\sqrt{2}$. Thus, the angle between $[124]_A$ and $[111]_B$ can be obtained from the following relation.

$$\left(\mathbf{a}_A + 2\mathbf{b}_A + 4\mathbf{c}_A\right)\cdot(\sqrt{2}\mathbf{b}_B + \mathbf{c}_B) = 2\sqrt{2}a_A a_B + 4c_A a_B = \sqrt{5a_A^2 + 16c_A^2}\sqrt{3a_B^2} \times \cos\theta$$

where $\theta = 35.04°$.

Similarly, $(111)_B$ is expressed as $(0\sqrt{2}1)$ in terms of the x and y coordinates. The interplanar angle of $37.79°$ is then obtained from the following relation.

$$\left(2\mathbf{a}_A^* + \mathbf{b}_A^* + \mathbf{c}_A^*\right)\cdot(\sqrt{2}\mathbf{b}_B^* + \mathbf{c}_B^*) = \frac{\sqrt{2}}{a_A a_B} + \frac{1}{c_A a_B} = \sqrt{\frac{5}{a_A^2} + \frac{1}{c_A^2}}\sqrt{\frac{3}{a_B^2}} \times \cos\theta$$

EXAMPLE 2.6.

A thin film of simple hexagonal lattice was deposited on a substrate of FCC lattice. Both materials are single-crystalline and are lattice-matched at the interface that is parallel to $(111)_{Sub}$ and $(001)_{Film}$. The substrate has a lattice constant of $a = \sqrt{2}$ Å and the film has $a = 1$ Å and $c = \sqrt{8/3}$ Å.

(a) Calculate the angle between $[100]_{Sub}$ and $[111]_{Film}$

(b) What is the angle between $(211)_{Sub}$ and $(120)_{Film}$?

Geometry of Crystals 65

Answer:

(a) A clue to solving this problem lies in the conversion of coordinates. Let **a**, **b**, and **c** be the unit cell vectors of the substrate. Since (001) of the film is lattice-matched to (111) of the substrate, the unit cell of the film can be drawn on $(111)_{Sub}$ as shown in Figure 2.33. If the unit cell vectors of the film are represented by $\mathbf{a}_F$, $\mathbf{b}_F$, and $\mathbf{c}_F$, we have the relations of $\mathbf{a}_F = \frac{1}{2}\mathbf{a} - \frac{1}{2}\mathbf{c}$ and $\mathbf{b}_F = -\frac{1}{2}\mathbf{a} + \frac{1}{2}\mathbf{b}$. The magnitude of $\mathbf{c}_F$ is $\sqrt{8/3}$ times the magnitude of $\mathbf{a}_F$ and the magnitude of $\mathbf{a}_F$ is $1/\sqrt{2}$ times the magnitude of **a**. This leads to $\mathbf{c}_F = \frac{2}{3}(\mathbf{a}+\mathbf{b}+\mathbf{c})$ because $\mathbf{c}_F$ is parallel to $[111]_{Sub}$, i.e., $\mathbf{a}+\mathbf{b}+\mathbf{c}$. Then, the vector representing [111] of the film is given by $\mathbf{r}_{111} = \mathbf{a}_F + \mathbf{b}_F + \mathbf{c}_F = \frac{2}{3}\mathbf{a} + \frac{7}{6}\mathbf{b} + \frac{1}{6}\mathbf{c}$. The angle between $[100]_{Sub}$ and $[111]_{Film}$ is thus equal to the angle between $[100]_{Sub}$ and $[271]_{Sub}$, which is 60.5°.

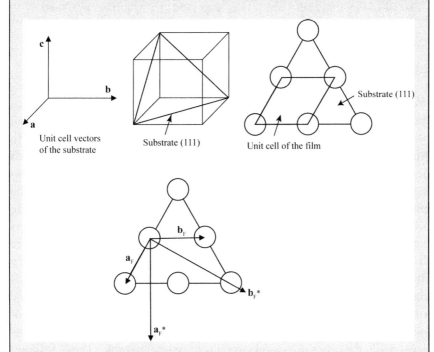

FIGURE 2.33 Orientation relationship between the substrate and the film.

(b) The angle between any two nonparallel planes is equal to the angle between their normal directions. Thus, the angle between the given two planes can be calculated with their reciprocal lattice vectors. The reciprocal lattice vector perpendicular to $(120)_{Film}$ is $\mathbf{H}_{120} = \mathbf{a}_F^* + 2\mathbf{b}_F^*$. With reference to Figure 2.33, $\mathbf{a}_F^*$ is parallel to $2\mathbf{a}_F + \mathbf{b}_F$ and $\mathbf{b}_F^*$ is parallel to $2\mathbf{a}_F + \mathbf{b}_F$. Then we have $\mathbf{H}_{120} = \mathbf{a}_F^* + 2\mathbf{b}_F^* = 4\mathbf{a}_F + 5\mathbf{b}_F = -\frac{1}{2}\mathbf{a} + \frac{5}{2}\mathbf{b} - 2\mathbf{c}$. Since we are concerned with the directions of the reciprocal lattice vectors, not their absolute magnitudes, the common proportionality constant can be neglected. The reciprocal lattice vector corresponding to $(211)_{Sub}$ is $\mathbf{H}_{211} = 2\mathbf{a}^* + \mathbf{b}^* + \mathbf{c}^*$. Here, $\mathbf{a}^*$, $\mathbf{b}^*$, and $\mathbf{c}^*$, all of an equal length, are parallel to $\mathbf{a}$, $\mathbf{b}$, and $\mathbf{c}$, respectively. Therefore, the corresponding reciprocal lattice vector can be simply set as $\mathbf{H}_{211} = 2\mathbf{a} + \mathbf{b} + \mathbf{c}$. Now, both of the reciprocal lattice vectors are represented in terms of the unit cell vectors $\mathbf{a}$, $\mathbf{b}$, and $\mathbf{c}$ of the substrate. The angle between $(211)_{Sub}$ and $(120)_{Film}$ is then equal to the angle between $[211]_{Sub}$ and $[\bar{1}5\bar{4}]_{Sub}$, which is calculated to be $86.4°$.

2.6 CRYSTAL STRUCTURES

We now consider the crystal structures of some important materials. Since the atoms in a crystal are periodically arranged, the crystal structure can be described by stating the relative positions of atoms within the unit cell. An element or compound may exhibit multiple crystal structures, each being the thermodynamically stable form in a given range of temperature and pressure. Only the outmost electrons are involved in binding the atoms together in a crystal and most of the electrons reside in the same orbits as in an isolated atom. Therefore, the atoms are usually regarded as hard spheres in describing the crystal structure. The simplest structures are those formed by placing atoms of the same kind on the lattice points of a Bravais lattice. However, so many crystals have more than one atom associated with each lattice point and the associated atoms may be of the same kind or different kinds. The set of atoms associated with each lattice point are termed the *basis* of the lattice. There are only fourteen Bravais lattices, while the types of crystal structures are innumerable. Thus, different crystal structures may have the same Bravais lattice. A vast majority of the elements are metallic. Most metals have one of the three structures: face-centered cubic (FCC), hexagonal close-packed (HCP), and body-centered

Geometry of Crystals 67

cubic (BCC). FCC and HCP are close-packed structures. FCC is the acronym for the face-centered cubic. The FCC structure should be differentiated from the FCC lattice. The former represents a specific crystal structure, while the latter is a lattice type that can be possessed by many different crystal structures. For instance, Au, Si, and NaCl have completely different structure but they have the same Bravais lattice of FCC. The primary chemical bonds between atoms can be divided into ionic, covalent, and metallic bonds, depending on the nature of inter-atomic binding. Since the metallic bond is non-directional, metal atoms tend to make as many bonds as possible with others. Accordingly, more than two-thirds of the metallic elements have either FCC or HCP structure. As will be discussed later in this section, the structures of many compounds can be derived from a close-packed structure. We start with common metallic structures.

2.6.1 FACE-CENTERED CUBIC STRUCTURE

Close-packed structures can be visualized as layers of hard spheres packed to maximum density both within layers and between adjacent layers. On a single layer, spheres of equal size are hexagonally arranged around a central sphere that touches all six neighbors. The spheres of the next layer should be placed in the grooves formed by three touching spheres to construct a close-packed structure, but only every other groove can be filled because the inter-groove distance is smaller than the diameter of the spheres. In Figure 2.34, the spheres of the first layer are located at points marked A (this layer is called A layer) and the second layer occupies positions marked B. The projections of the spheres centered on points B are shown shaded in the figure. There are two possibilities for the third layer because the second layer has two different types of groove positions: C and A. When the spheres of the third layer are stacked at the C positions, the fourth layer should be located at positions A to meet the intrinsic translational symmetry of the crystal. The resulting structure is FCC, which has the stacking sequence of $ABCABC...$

68   X-Ray Diffraction for Materials Research: From Fundamentals to Applications

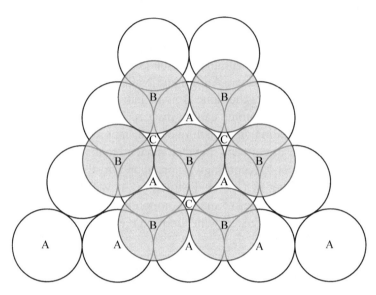

FIGURE 2.34 Close packing of hard spheres.

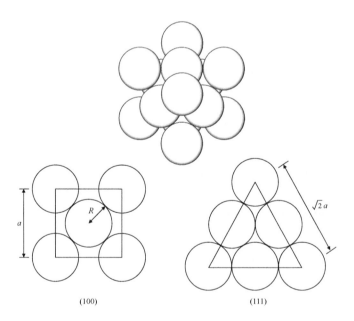

FIGURE 2.35 Face-centered cubic (FCC) structure.

Figure 2.35 shows a conventional unit cell of the FCC structure, along with the atomic arrangements on (100) and (111) planes. The radius of the atom, R, and the lattice parameter, a, have the relation of $R = a/2\sqrt{2}$. The {111} planes are the close-packed layers in FCC, which are stacked in the sequence of $ABCABC...$ along the <111> directions. Figure 2.36(a) shows successive (111) planes. Identical atoms are differently shaded on different planes for distinction. The stacking sequence can be easily figured out by examining the atomic arrangement on ($1\bar{1}0$), which is perpendicular to the (111) planes. The ($1\bar{1}0$) plane is represented as a bold-line rectangle in Figure 2.36(b). A line connecting the lower-left corner of this rectangle to the upper-right corner is parallel to [111]. When viewed along the [111] direction, the first (111) plane has atoms at the A position. The second and third planes have atoms on the B and C positions, respectively. The atoms of the fourth plane are coincident with those of the first plane. Thus, we have a stacking sequence of $ABCABC$. This structure has triads along the four body-diagonal directions of the unit cell. Since all atoms have identical environments, each atom constitutes one lattice point. The Bravais lattice of the FCC structure is also FCC. Conversely, this structure is obtained when equal atoms are placed at the lattice points of an FCC lattice (see Figure 2.3). There are 4 atoms per unit cell: one at the corner and three at the face centers. The coordinates of the atoms are (0,0,0), (1/2,1/2,0), (1/2,0,1/2), and (0,1/2,1/2). Although a single unit cell has eight corner atoms and six face-centered atoms, the corner atom is shared by eight unit cells and the face-centered one, by two cells. Each atom has twelve nearest neighbors, that is, the coordination number is 12.

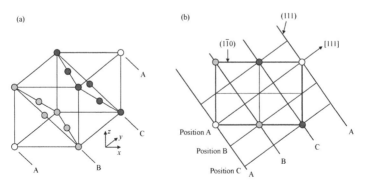

FIGURE 2.36 (a) Successive (111) planes in FCC structure. (b) Atomic arrangement on ($1\bar{1}0$). Identcal atoms are differently shaded for distinction.

When the structure is regarded as consisting of spheres in contact, the interstices between spheres are important because many compound structures contain one set of atoms in an FCC arrangement, with the others occupying the interstices. The largest interstices are located at coordinate (1/2,1/2,1/2) and equivalent positions (0,1/2,0; 0,0,1/2; 1/2,0,0). There are four such interstices per unit cell. Since this interstice has octahedral coordination with six neighbors (Figure 2.37), its center is called *octahedral site*. The octahedral site is located midway between two adjacent (111) planes. The largest sphere that can go into the octahedral site of the FCC structure without lattice distortion has radius $r = (\sqrt{2} - 1)R = 0.414R$. The second largest interstices are located at (1/4,1/4,1/4) and equivalent positions. Since this interstice is formed by four atoms that possess tetrahedral coordination, it is called *tetrahedral site*. The largest sphere that can enter the tetrahedral site has radius $r = (\sqrt{3/2} - 1)R = 0.225R$. There are eight tetrahedral sites in the unit cell. When viewed along the <111> directions, four of them are positioned at the centers of upright tetrahedrons and the other four, at the centers of inverted tetrahedrons (Figure 2.38). It should be noted that the tetrahedral sites are not located midway between two (111) planes. Their positions are $d_{111}/4$ or $3d_{111}/4$ away normally from the close-packed layers, depending on the upright or inverted type. Many metals including Ag, Au, Al, Ni, Co, Ni, and γ-Fe have this FCC structure.

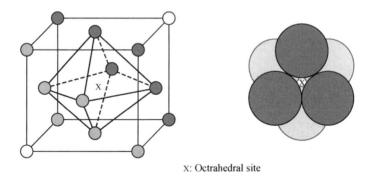

X: Octahedral site

FIGURE 2.37 Octahedral site in FCC structure. The octahedral site is located midway between two adjacent (111) planes and has six neighboring atoms.

Geometry of Crystals

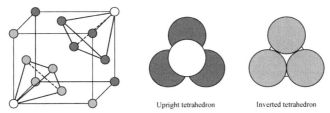

FIGURE 2.38 Tetrahedral sites in FCC are at the centers of regular tetrahedrons, each formed by four atoms.

EXAMPLE 2.7. NUMBER OF BROKEN BONDS VS. SURFACE ENERGY

An FCC crystal with lattice parameter $a = 1$ nm was cut and polished so that its surfaces consist of (100), (110), and (111) planes. Calculate the number of broken bonds per unit area on each surface plane and state which one has the highest surface energy.

Answer:

The coordination number of the FCC structure is 12, that is, each atom makes bonds with 12 neighboring atoms. As shown in Figure 2.35, any atom on the (100) surface makes bonds with 4 atoms on the same surface. Since it has 4 other bonds with atoms residing just below the surface, the number of broken bonds per atom is 4. An atom on the (111) surface makes 6 bonds on the same surface. Therefore, the number of broken bonds is 3 because it has bonding with 3 atoms below the surface.

An atom of the (110) surface has only 2 bonds on the same surface, as shown in Figure 2.39. Thus, the number of broken bonds per atom is 5. In all cases, the number of broken bonds per atom is obtained by subtracting the number of surface bonds from the coordination number 12 and dividing the remainder by two. The number of broken bonds per unit area on each surface plane can be calculated by combining the number of broken bonds per atom with the number of atoms per unit surface area. The results are given in Table 2.3. The surface energy arises from the broken bonds and is proportional to the number of broken bonds per unit area. Therefore, the (100) surface has the highest surface energy.

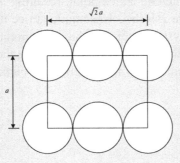

FIGURE 2.39 Atomic arrangements on the (110) plane.

72   X-Ray Diffraction for Materials Research: From Fundamentals to Applications

TABLE 2.3   Number of Broken Bonds Per Unit Area On the Different Surfaces of An FCC Crystal With $a = 1$ nm

	(100) surface	(110) surface	(111) surface
Number of broken bonds per atom	4	5	3
Number of surface atoms per unit area	$\frac{2}{a^2} = \frac{2}{\text{nm}^2}$	$\frac{2}{(\sqrt{2}a^2)} = \frac{1.414}{\text{nm}^2}$	$\frac{4}{(\sqrt{3}a^2)} = \frac{2.309}{\text{nm}^2}$
Number of broken bonds per unit area	8/nm^2	7.07/nm^2	6.927/nm^2

2.6.2 HEXAGONAL CLOSE-PACKED STRUCTURE

Let's go back to Figure 2.34. There are two ways of stacking the third layer because the second layer has two different types of groove positions: C and A. When the spheres of the third layer are stacked at the A positions, the close packing is also maintained. If the stacking proceeds in this sequence of $ABAB...$, we obtain the HCP structure. The unit cell contains two identical atoms with coordinates $(0,0,0)$ and $(2/3,1/3,1/2)$, as illustrated in Figure 2.40(a). The atoms on (002) planes of the HCP structure are arranged in a hexagonal pattern just like those on the $\{111\}$ planes of the FCC structure. The only difference between the two structures is the way in which these 2D close-packed layers are stacked above one another. The two atoms at $(0,0,0)$ and $(2/3,1/3,1/2)$ have different environments. Thus, they together constitute one lattice point, resulting in a simple hexagonal Bravais lattice (Figure 2.40(b)). The simple lattice is the only lattice type that the hexagonal crystal system can have. Sometimes, the crystal structure can be more easily figured out by projecting the atomic centers on a specific plane rather than drawing an actual 3D picture.

Geometry of Crystals 73

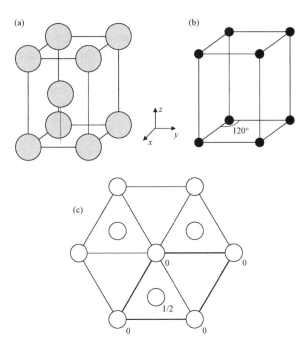

FIGURE 2.40 (a) Hexagonal close packed (HCP) structure; (b) Simple hexagonal Bravais lattice; (c) (001)-projection.

In Figure 2.40(c), the atomic centers are projected on the (001) plane of the HCP structure. The numerical values given on this (001)-projection represent the relative heights of atoms within the unit cell. The octahedral sites have coordinates (1/3,2/3,1/4) and (1/3,2/3,3/4). The tetrahedral sites lie at (0,0,3/8), (0,0,5/8), (2/3,1/3,1/8), and (2/3,1/3,7/8). Of course, two of them are the centers of upright tetrahedrons and the other two sites, the centers of inverted tetrahedrons. If each sphere has 12 nearest neighbors, the axial ratio of the unit cell, c/a, should be $\sqrt{8/3} = 1.633$. The packing fraction, the proportion of space filled by the spheres, is then 0.74, as in the FCC structure. A packing fraction of 0.74 is the highest value that can be achieved in element crystals. Many metals with this HCP structure (e.g., Zn, Co, Cd, Mg, Be, Ti at room temperature) have axial ratios more or less different from the ideal value of 1.633. It is attributed to the fact that the atoms in these crystals are ellipsoidal in shape rather than spherical. It may look at first that the HCP structure lacks six-fold rotational symmetry and possesses triads only. However, its crystal system is obviously hexagonal.

Consider two atomic planes (plane 1 and 2) that run normal to the basal plane of the HCP structure and that make an angle of 60° with each other. Figure 2.41 shows that these two planes have the same atomic arrangements. Namely, the two planes are crystallographically identical. This means that the HCP structure comes into self-coincidence after rotation of 60° about its **c** axis, revealing the presence of six-fold rotation symmetry.

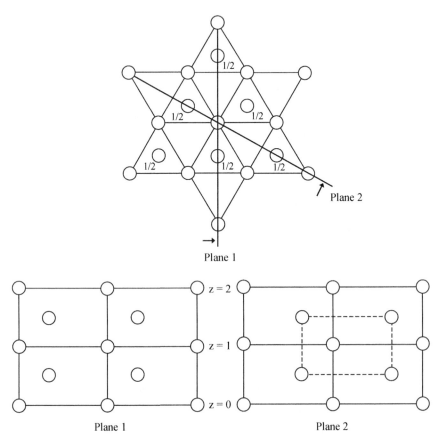

FIGURE 2.41 Atomic arrangements on two atomic planes perpendicular to the HCP basal plane. "z" represents fractional heights relative to the lattice parameter c.

2.6.3 BODY-CENTERED CUBIC STRUCTURE

This structure is represented by some alkali metals (Li, Na, K at room temperature) and transition metals (V, Cr, Mo, Nb, Ta, W, and α-Fe). There are two atoms per unit cell and the atomic coordinates are (0,0,0) and (1/2,1/2,1/2) (Figure 2.42). The Bravais lattice is BCC with one atom at each lattice point. Each atom has eight nearest neighbors. Unlike FCC and HCP, there are no closest packed planes in this structure. If the structure is constructed with equal spheres, then $R = \sqrt{3}a/4$. The packing fraction is 0.68. When made up of equal spheres, this structure has largest interstices at coordinates (1/2,1/4,0) and equivalent positions. There are twelve such interstices. The largest sphere that can fit in such an interstice without lattice distortion has radius equal to 0.228R. This interstice has tetrahedral coordination with four neighbors but the tetrahedron formed by these neighbors is not regular. The second largest interstices are at (1/2,1/2,0) and equivalent positions. There are six such interstices per unit cell (three at the face centers and three at the midpoints of cube edges). This site is located at the center of a distorted octahedron and can accommodate a sphere of radius $r = 0.15\,R$. Metals that have a BCC structure are usually harder and less malleable than close-packed metals such as gold and silver. Slip is an important mechanism of plastic deformation in metals. When the metal is deformed, the planes of atoms must slip over each other. The slip planes are normally the planes with the highest atomic density, i.e., those most widely spaced. Slip in FCC crystals occurs along the close-packed {111} planes. Unlike FCC, there are no truly close-packed planes in the BCC structure. This is one of the reasons why BCC metals are more difficult to plastically deform than FCC metals.

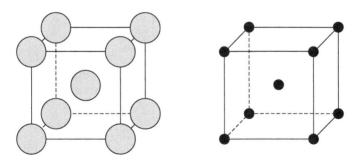

FIGURE 2.42 Body-centered cubic (BCC) structure and its lattice.

EXAMPLE 2.8. RECIPROCAL LATTICE OF FCC CRYSTAL

Prove that the reciprocal lattice of an FCC structure with lattice constant "a" is BCC with lattice constant $2/a$.

Answer:

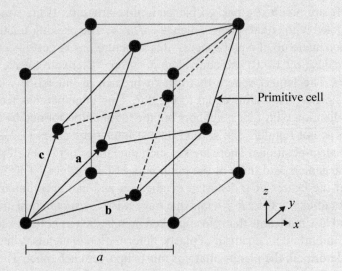

FIGURE 2.43   A primitive cell of the FCC lattice.

The FCC structure has an FCC lattice, as shown in Figure 2.43. The unit cell of the reciprocal lattice has been derived using the primitive cell vectors of the real lattice (see Figure 2.26 and Eq. (2.14)). To construct the reciprocal lattice of this FCC lattice, we need to start from its primitive cell vectors, which are expressed as

$$\mathbf{a} = \mathbf{i}\frac{a}{2} + \mathbf{k}\frac{a}{2}, \mathbf{b} = \mathbf{i}\frac{a}{2} + \mathbf{j}\frac{a}{2}, \mathbf{c} = \mathbf{j}\frac{a}{2} + \mathbf{k}\frac{a}{2}$$

where **i**, **j**, and **k** are unit vectors in the x, y, and z directions of Figure 2.43. The vector products of these three vectors are

$$\mathbf{a} \times \mathbf{b} = a^2 \begin{vmatrix} \mathbf{i} & \mathbf{j} & \mathbf{k} \\ \frac{1}{2} & 0 & \frac{1}{2} \\ \frac{1}{2} & \frac{1}{2} & 0 \end{vmatrix} = -\mathbf{i}\frac{a^2}{4} + \mathbf{j}\frac{a^2}{4} + \mathbf{k}\frac{a^2}{4}$$

Geometry of Crystals

$$\mathbf{b} \times \mathbf{c} = a^2 \begin{vmatrix} \mathbf{i} & \mathbf{j} & \mathbf{k} \\ 1/2 & 1/2 & 0 \\ 0 & 1/2 & 1/2 \end{vmatrix} = \mathbf{i}\,a^2/4 - \mathbf{j}\,a^2/4 + \mathbf{k}\,a^2/4$$

$$\mathbf{c} \times \mathbf{a} = a^2 \begin{vmatrix} \mathbf{i} & \mathbf{j} & \mathbf{k} \\ 0 & 1/2 & 1/2 \\ 1/2 & 0 & 1/2 \end{vmatrix} = \mathbf{i}\,a^2/4 + \mathbf{j}\,a^2/4 - \mathbf{k}\,a^2/4$$

Since the volume of the primitive cell is $V = a^3/4$, we obtain the following relations for the reciprocal lattice vectors.

$$\mathbf{c}^* = \frac{\mathbf{a} \times \mathbf{b}}{V} = -\mathbf{i}/a + \mathbf{j}/a + \mathbf{k}/a$$

$$\mathbf{b}^* = \frac{\mathbf{c} \times \mathbf{a}}{V} = \mathbf{i}/a + \mathbf{j}/a - \mathbf{k}/a$$

$$\mathbf{a}^* = \frac{\mathbf{b} \times \mathbf{c}}{V} = \mathbf{i}/a - \mathbf{j}/a + \mathbf{k}/a$$

These three vectors are the vectors pointing to the body centers of adjacent three cubes from the origin. That is, they form the primitive unit cell vectors of a BCC lattice with lattice parameter $2/a$, as shown in Figure 2.44.

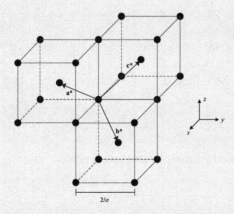

FIGURE 2.44 Primitive unit cell vectors of a BCC lattice.

2.6.4 DIAMOND STRUCTURE

The diamond structure can be described with an FCC lattice coupled with a basis of two identical atoms. Its Bravais lattice is FCC and two atoms are associated with one lattice point. For instance, two identical atoms placed at (0,0,0) and (1/4,1/4,1/4) form the lattice point at (0,0,0). Not only the FCC structure but also many other structures have an FCC lattice. A conventional unit cell of the diamond structure contains eight atoms. When four atoms are arranged in an FCC fashion (i.e., located at the corners and face centers of the cube), the other four occupy half of the tetrahedral sites. Figure 2.45 shows the diamond structure and its (001)-projection. In this configuration, each atom lies at the center of a regular tetrahedron formed by four nearby atoms. Note that the corner atom with coordinate (0,0,0) is also located at the center of a tetrahedron made with four atoms at (1/4,1/4,1/4), (−1/4,1/4,−1/4), (1/4,−1/4,−1/4), and (−1/4,−1/4,1/4). Accordingly, the coordination number of each atom is four. The stacking sequence of successive (111) planes can be described as *A AB BC CA AB BC CA*. Successive planes are not equally separated from one another. This is an open structure with the packing fraction of 0.34. Si, Ge, and diamond (i.e., crystalline carbon stable at high temperature and pressure) have this structure.

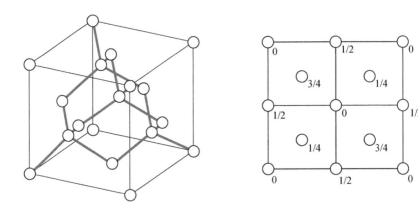

FIGURE 2.45 Diamond structure and (001)-projection.

2.6.5 NaCl AND NiAs STRUCTURES

A third of the type *MX* compounds exhibit the NaCl structure shown in Figure 2.46(a). Its Bravais lattice is FCC with two different atoms associated with one lattice point. For example, M at $(0,0,0)$ and X at $(0,0,1/2)$ constitute the lattice point at $(0,0,0)$. This structure has a configuration where one type of atoms (or ions) is located at the lattice points and atoms of the other type occupy all of the octahedral sites. In this structure, each atom is at the center of a regular octahedron formed by atoms of the other type. The coordination number is thus six. There are four formula units per conventional unit cell. Each {111} lattice plane specifies two sheets of atoms, each sheet consisting of atoms of the same type. If we denote atoms of one kind with Roman letters and those of the other with Greek letters, the stacking sequence along the [111] direction can be described as $A\gamma B\alpha C\beta A\gamma B\alpha C\beta A$. Many oxides (BaO, CaO, MgO, FeO, etc.) and alkali halides (LiCl, LiBr, KCl, KBr, etc.) have this NaCl structure. The NaCl structure can be derived from the FCC structure by placing another set of atoms in its octahedral sites. It is not surprising that another structure may be similarly obtained, in which atoms of a different kind are placed in the octahedral sites of the HCP structure. This is the NiAs structure shown in Figure 2.46(b). The Bravais lattice is simple hexagonal. There are four atoms in the unit cell: one kind at $(0,0,0)$ and $(2/3,1/3,1/2)$ and the other kind at $(1/3,2/3,1/4)$ and $(1/3,2/3,3/4)$. The stacking sequence of the atomic planes along [001] is $A\gamma B\gamma A\gamma B\gamma A$. FeS, CoS, NiS, and CrS have this NiAs structure.

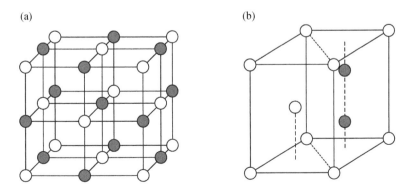

FIGURE 2.46 (a) NaCl structure. (b) NiAs structure.

2.6.6 SPHALERITE AND WURTZITE STRUCTURES

The FCC structure has four atoms and eight tetrahedral sites per conventional unit cell. Thus, for compounds of the type *MX*, only half of the tetrahedral sites should be filled. The resulting structure is the sphalerite structure in which four atoms of one kind are at the unit cell corners and face centers and four atoms of a different kind occupy every other tetrahedral site (Figure 2.47(a)). The sphalerite structure is sometimes called the zinc blende structure. The Bravais lattice is also FCC with two different atoms associated with one lattice point. The coordination number is four. When all atoms are of the same kind, it renders to the diamond structure. The stacking sequence of (111) planes is $A \ \alpha B \ \beta C \ \gamma A \ \alpha B \ \beta C \ \gamma A$. A number of compounds exhibit this structure, which include GaAs, α-ZnS, InP, and InAs.

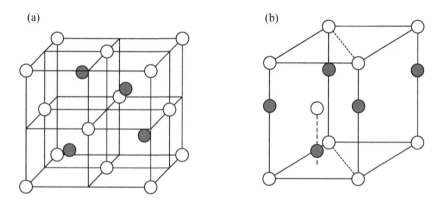

FIGURE 2.47 (a) Sphalerite structure. (b) Wurtzite structure.

The hexagonal version of the sphalerite structure is the wurtzite structure that is obtained by filling alternate tetrahedral sites in the HCP structure (Figure 2.47(b)). There are two atoms of one kind at (0,0,0) and (2/3,1/3,1/2) and two atoms of another kind at (0,0,5/8) and (2/3,1/3,1/8). Each atom is of course tetrahedrally coordinated with four atoms of the opposite kind. The Bravais lattice is simple hexagonal and a total of four atoms is thus associated with one lattice point. The stacking sequence along [001] is $A \ \alpha B \ \beta A \ \alpha B \ \beta A$... or equivalently $A\beta \ B\alpha \ A\beta \ B\alpha \ A$. ZnO, β-ZnS, SiC, and GaN have this wurtzite structure. While the stacking se-

Geometry of Crystals

quence can be easily figured out in the hexagonal crystals, it is not immediately obvious in the case of cubic crystals. Figure 2.48 illustrates how the NaCl and sphalerite structures have the given sequences. It is to be noted that successive planes are equally separated from one another in the NaCl structure but not in the sphalerite structure.

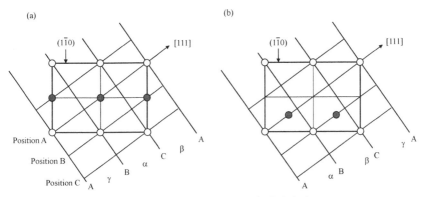

FIGURE 2.48 Stacking sequences in (a) NaCl and (b) Sphalerite structures.

2.6.7 OTHER COMPOUNDS

Figure 2.49 shows two more compound structures: CsCl and perovskite structures. Many intermetallic compounds such as FeAl, CuZn, and AgMg have the CsCl structure. In this structure, one atom is located at the unit cell corner and its body center is filled with another atom of a different kind. Since these two atoms have different environments, they together form a single lattice point. For instance, Cs at (0,0,0) and Cl at (1/2,1/2,1/2) are associated with the lattice point at (0,0,0). Accordingly, the Bravais lattice of the CsCl structure is simple cubic. The *perovskite structure* is a common structure possessed by a number of compounds of the type $MM'X_3$, which include $CaTiO_3$, $SrTiO_3$, and $BaTiO_3$. The corners and body center of the unit cell are filled with the M and M' atoms (or ions), respectively, and the X atoms are positioned at its face centers. In $CaTiO_3$, the atomic coordinates are Ca^{2+} (0,0,0), Ti^{4+} (1/2,1/2,1/2), and O^{2-} (1/2,1/2,0), (1/2,0,1/2), (0,1/2,1/2). Many perovskite compounds exhibit phase transitions. For example, $BaTiO_3$ is cubic above 120°C, tetragonal in the temperature range of 120° to 7°C, and transforms to an orthorhombic phase below 7°C. In a strict sense, it maintains the perovskite structure only above 120°C.

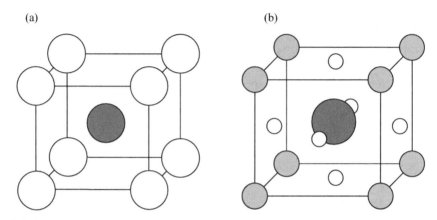

FIGURE 2.49 (a) CsCl structure. (b) Perovskite structure.

If all of the tetrahedral sites in the FCC structure are occupied with atoms of a different kind, we obtain the CaF_2 structure. The Bravais lattice of this structure is also FCC. Each lattice point is associated with one Ca atom and two F atoms. One unit cell contains four formula units. When Ca atoms are located at the lattice points of an FCC lattice, F atoms occupy all of its tetrahedral sites. The coordination number of Ca is eight and F has tetrahedral coordination with four Ca atoms. More complex structures are also derived from the FCC structure. The *spinel structure*, possessed by $MgAl_2O_4$ and other mixed oxides of di-and trivalent metals, has a unit cell containing 32 oxygen ions arranged in cubic close packing. Mg^{2+} ions occupy one-eighth of the 64 tetrahedral sites and Al^{3+} occupy half of the 32 octahedral sites. *Inverse spinel structures* have a different cation distribution. $MgFe_2O_4$ is an example of the inverse spinel structure. Here, the sixteen octahedral sites are filled by all of the Mg ions and half of the Fe ions, while the eight tetrahedral sites are filled with the remaining Fe ions.

2.6.8 SOLID SOLUTIONS

Many pure metals can dissolve other elements to form solid solutions. There are two types of solid solutions: substitutional and interstitial solid solutions. In the former, solute atoms merely substitute for the solvent atoms. In the latter, the solute atoms fit into the interstices of the solvent atoms. Unless the size difference is so large, the solute and solvent atoms

Geometry of Crystals 83

are likely to form a substitutional solution. Let's consider a binary alloy system consisting of *A* and *B* elements, in which pure *A* has a BCC structure and pure *B*, an FCC structure (Figure 2.50). Any metal can dissolve other elements to a degree. Thus, when the *A* element contains a small amount of *B* solute atoms, it maintains a single phase denoted as α phase. This is equally applied to the *B* element. If the concentration of the solute atoms is above a threshold value, *A*-rich α phase and *B*-rich β phase coexist because the two elements have different structures and a single phase cannot be maintained in intermediate compositions. In solid solutions, the solute atoms are somewhat randomly distributed. Therefore, the *B* atoms in the α phase may occupy either the corners or body centers of the unit cell in an irregular manner, as shown in Figure 2.50. As a result, some unit cells will not exhibit cubic symmetry. Nevertheless, the overall structure of the α phase is still BCC. The X-ray beam used to examine the crystal is much larger than the size of a unit cell. A tremendous number of unit cells are analyzed at the same time. We take only an average picture of the structure. Likewise, the β phase has the same FCC structure as the pure *B* element. In an intermediate composition, the alloy may consist of two phases. In this case, the resulting X-ray diffraction pattern will be a mixture of the characteristic patterns of BCC and FCC structures.

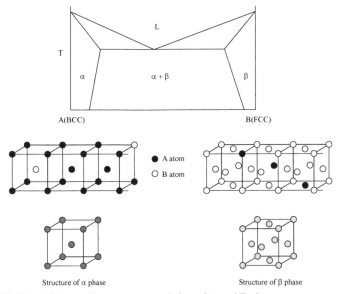

FIGURE 2.50 A binary alloy system consisting of **A** and **B** elements.

It is noteworthy of a bit mentioning the ordered and disordered structures. Generally, substitutional solid solutions have solute atoms randomly distributed among the available sites. In some cases, this random distribution is true only at high temperatures and ordering occurs below a certain temperature. The intermetallic compound $AuCu_3$ is a classic example. Above 390°C, both atoms are randomly positioned at the corners and face centers of the cubic unit cell (Figure 2.51). There is no preferred position for Cu or Au. The probability that a particular atomic site is occupied by Au is 1/4, which is the atomic fraction of Au in the alloy. Accordingly, the probability that the same site is occupied by Cu is 3/4. Since every site is identical in terms of the occupation probability, the structure of this disordered state can be regarded as an FCC. Its Bravais lattice is thus FCC. When cooled down below 390°C, the Au atoms occupy the corners of the unit cell and the Cu atoms, the face centers. While the disordered state has an FCC lattice, this ordered structure has a simple cubic lattice. Since the diffraction pattern depends on the lattice type, we can determine the order-disorder transition temperature by X-ray diffraction.

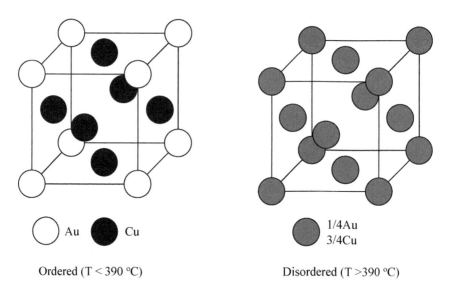

FIGURE 2.51 Ordered and disordered states in Cu_3Au.

Geometry of Crystals

EXAMPLE 2.9

The Ni$_2$In structure has a hexagonal unit cell with atoms in the following positions:

Ni: (0,0,0) (0,0,1/2) (2/3,1/3,3/4) (1/3,2/3,1/4)

In: (2/3,1/3,1/4) (1/3,2/3,3/4)

(a) Draw the Ni$_2$In structure and represent the (001)-projection.

(b) Draw the atomic arrangements in an alternative unit cell with **In** atoms on the corners.

Answer:

Figure 2.52(a) shows the Ni$_2$In structure and its (001)-projection drawn with the given atomic coordinates. An alternative unit cell with **In** atoms on the corners can be taken, as shown in Figure 2.52(b).

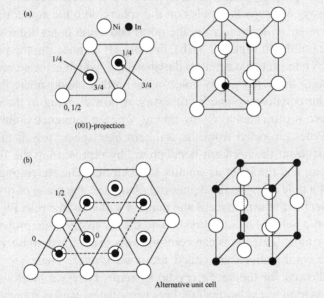

FIGURE 2.52 (a) Ni$_2$In structure and its (001)-projection drawn with the given atomic coordinates. (b) An alternative unit cell with **In** atoms on the corners.

2.7 STEREOGRAPHIC PROJECTION

When we study crystallography and crystal structures, it is often necessary to represent planes or directions on a two-dimensional diagram so that the crystal symmetry and angular relationships can be easily figured out. The orientation of any plane in a crystal can be represented by the inclination of its normal relative to a reference plane. Suppose that a crystal is positioned at the center of a sphere, called the *sphere of projection*, as shown in Figure 2.53(a). When the normal of a plane is drawn from the sphere center to intersect the surface of the sphere at P, the point P is called the *pole* of the plane. The orientation of a plane is thus represented by a pole on the sphere and the line OP is normal to the plane. To represent the crystal planes on a two-dimensional diagram, the poles should be projected on to a plane, denoted as the *plane of projection*. In analogy with the earth, let us define points N and S as the north and south poles. Then, it is very convenient if the equatorial plane is selected as the plane of projection. There are various ways of projecting poles on the sphere onto the equatorial plane. In the *stereographic projection*, the north and south poles are used as the reference points. In Figure 2.53(b), the line SP_1 connecting the pole P_1 to the point S intersects the equatorial plane at P_1', which is the stereographic projection of the pole P_1. Any poles in the northern hemisphere fall inside the circular equatorial plane by this way. A pole P_2 lying in the southern hemisphere is projected to P_2' by taking N as the reference point for projection. Poles projected from the southern hemisphere are distinguished from those from the northern hemisphere, by representing the former as open circles and the latter as smaller solid circles. The stereographic projection of a pole lying on the equatorial plane, i.e., the plane of projection, is consistent with the pole itself and marked with a solid circle. Figure 2.54 shows some poles of a cubic crystal and their stereographic projection.

The symmetry elements can combine only in a limited number of ways and these combinations are called the *point groups*. There are 32 point groups allowed for the seven crystal systems. Here we briefly examine how the symmetric relationships between planes can be represented on a stereogram in the tetragonal case. When the crystal has a tetrad only, its point group is denoted as "4" (Figure 2.55). If there is a mirror plane perpendicular to the tetrad, the crystal has a point group "4/m". When the crystal possesses mirror planes parallel to the four-fold axis as well, the point group becomes "4/mmm".

Geometry of Crystals

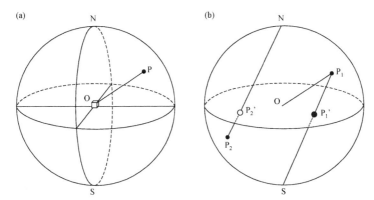

FIGURE 2.53 (a) Sphere of projection. (b) Stereographic projection.

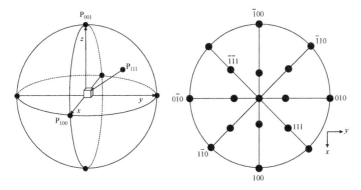

FIGURE 2.54 (a) Poles of a cubic crystal. (b) Stereographic projection of poles.

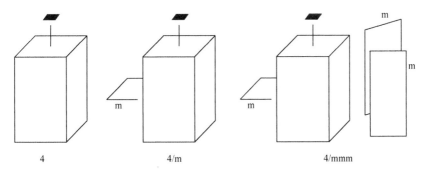

FIGURE 2.55 Point groups of the tetragonal system and their symmetry elements.

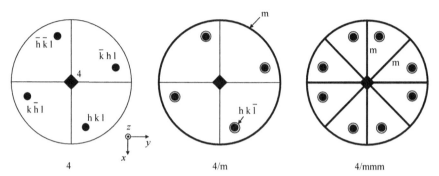

FIGURE 2.56 Stereograms of three different point groups. The characteristic four-fold axis is parallel to the z axis.

Figure 2.56 shows the stereograms of three different cases. The operation of four-fold rotational symmetry to a general pole (*hkl*) produces three equivalent poles with indices ($\bar{k}hl$), ($\bar{h}\bar{k}l$), and ($k\bar{h}l$). When there is a mirror plane running perpendicular to the four-fold rotation axis, four more equivalent poles are generated. Now, (*hkl*) and ($\bar{h}\bar{k}\bar{l}$) planes are crystallographically identical and thus, the crystal of point group 4/m possesses a center of symmetry. In the point group 4/mmm, the number of planes equivalent to (*hkl*) increases to fifteen: sixteen including itself. It exhibits the highest symmetry available in the tetragonal system. As stated in Section 2.6.7, BaTiO$_3$ is cubic above 120°C. When cooled through 120°C, it transforms to a tetragonal phase and exhibits a dipole moment. The dipole moment is considered to arise primarily due to the movement of Ti ions with respect to the O ions in the same plane, but the movement of the other O ions (i.e., those above and below Ti ions) and the Ba ions is also relevant. Figure 2.57 illustrates a simplified mechanism of the cubic-to-tetragonal transformation. In the cubic phase, BaTiO$_3$ has three orthogonal tetrads along the unit cell axes and mirror planes parallel as well as perpendicular to these tetrads. When the phase transition occurs, only the tetrad parallel to the direction of ions movement remains. This direction is defined as the *z*-axis direction of the tetragonal unit cell. A mirror plane perpendicular to the tetrad no longer exists but those parallel to it are still maintained. This is why the tetragonal BaTiO$_3$ has a point group "4mm". A center of symmetry is absent from this point group. 4/m and 4mm have the same number of equivalent poles but their indices are different.

Geometry of Crystals

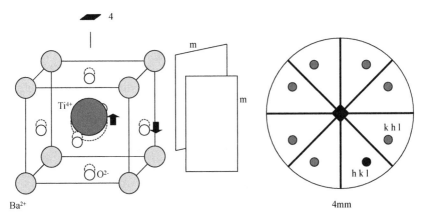

FIGURE 2.57 A simplified mechanism of the cubic-to-tetragonal transformation in $BaTiO_3$. In the tetragonal phase, $BaTiO_3$ has a point group of 4mm.

PROBLEMS

2.1. Determine the angle between [111] and the normal to (111) in a tetragonal crystal with $a = 5$ Å and $c = 12$ Å.

2.2. Prove that the reciprocal lattice of a BCC structure is FCC. If this BCC structure has lattice constant "a", what is the volume of a primitive cell of the reciprocal lattice?

2.3. For a tetragonal crystal with 4/mmm point group, write down all planes that are crystallographically identical with (321). Which planes should be excluded from the form {321} if the crystal becomes 4mm due to a phase change?.

2.4. A certain tetragonal crystal has the following d-spacings:
$d_{002} = 4.68$ Å, $d_{101} = 3.47$ Å
Find the magnitudes of the lattice parameters a and c.

2.5. There is a monoclinic crystal with $a = 8.035$ Å, $b = 5.805$ Å, $c = 7.346$ Å, $\beta = 105.63°$. Calculate the reciprocal lattice constants and the spacing of (111) planes, i.e., d_{111}.

2.6. A crystal has unit cell with $a = 9.85$ Å, $b = 15.42$ Å, $c = 7.71$ Å, $\alpha = \beta = \gamma = 90°$.
(a) Represent (110), (123), and [211] in a unit cell.
(b) What is the angle between [111] and the normal to (123)?

(c) Let (134) and (213) belong to a specific zone axis. Then write down the indices of two more planes that belong to this zone axis.

2.7. A compound of the type ABC_3 has unit cell dimensions of $a = b = c$, $\alpha = \beta = \gamma = 90°$. The atomic coordinates within the unit cell are as follows.

A: (1/2,1/2,1/2) B: (0,0,0) C: (1/2,0,0), (0,1/2,0), (0,0,1/2)

(a) What are the crystal system and the Bravais lattice of this compound?

(b) Give the (001)-projection and draw the atomic arrangement in an alternative unit cell with "A" atoms on the corners.

2.8. Calculate the angle between (100) and (111) in each of the following metals.

(a) Cu: Cubic

(b) Sn: Tetragonal with $a = 5.82$ Å and $c = 3.17$ Å.

(c) Zn: Hexagonal with $a = 2.66$ Å and $c = 4.93$ Å.

2.9. In an AB compound, A ions are in contact with twelve A ions and six B ions. What is the radius ratio of the two ions and what kind of structure does this compound have?

2.10. Iron (Fe) has a BCC structure at room temperature but FCC is a stable structure above 910°C. When a block of iron is heated to over 910°C, how much volume change will occur to this block?

2.11. In a cubic crystal, the angle between a plane P and (111) is 53.96°, and P lies in the [010] zone.

(a) Find the Miller indices of the plane P.

(b) Find the angle between P and the plane P* if P* is related to P by a mirror plane parallel to $(1\bar{1}0)$.

2.12. In a tetragonal crystal, the angle between [111] and $(\bar{1}\bar{1}1)$ is 108.67°.

(a) Calculate the axial ratio, c/a.

(b) Calculate the angle between [001] and [236].

(c) Calculate the angle between (111) and (213).

2.13. In a cubic crystal, (hkl), (101), and (011) belong to a zone. This (hkl) plane also lies in another zone containing (213) and (211). What is (hkl) if the angle between (hkl) and (101) is 19.01°?

2.14. Calculate the atomic packing fraction of diamond structure.

Geometry of Crystals   91

2.15. In a hexagonal unit cell, indicate the following planes and directions: $(1\bar{2}10)$, $(10\bar{1}2)$, $(\bar{1}011)$, $[110]$, $[11\bar{1}]$, and $[021]$.

2.16. There is a hypothetical crystal ($a = b = c$, $\alpha = \beta = \gamma = 90°$) with the following atomic coordinates. Give the Bravais lattice for each case.

(a) A: (0,0,0) B: (0,0,1/2)

(b) A: (0,0,0) B: (1/2,0,1/4) C: (0,1/2,3/4)

(c) A: (0,0,0) B: (1/2,1/2,1/2) C: (1/2,1/2,0) (1/2,0,1/2) (0,1/2,1/2)

2.17. Diamond structure has hexagonal-shaped tunnels when it is viewed perpendicular to (110). Calculate the cross-sectional area of the tunnel.

2.18. State the definition of *center of symmetry* and its significance in relation to the material property.

2.19. β-ZnS is hexagonal with the following coordinates.

S: (0,0,0), (2/3,1/3,1/2) Zn: (0,0,3/8), (2/3,1/3,7/8)

(a) Represent the (001)-projection

(b) Draw the atomic arrangements in an alternative unit cell with Zn atoms on the corners.

(c) How does this β-ZnS structure (Wurtzite structure) differ from the α-ZnS structure.

2.20. In a material consisting of A and B elements, A atoms are arranged in an FCC fashion and B atoms occupy all of the tetrahedral and octahedral sites.

(a) What would be the stoichiometry of a compound with this structure?

(b) In the disordered state where the A and B atoms are randomly distributed among the available positions, is the structure equivalent to BCC?

2.21. There is a BCC crystal with $a = 1.0$ nm. Calculate the diameters of the largest atoms that can go into the tetrahedral and octahedral sites without lattice distortion.

2.22. Identify all planes identical to (123) in the point group 4/mmm.

2.23. Calculate the number of broken bonds per atom on the (100) and (110) surfaces of diamond structure.

2.24. Determine the angle between (121) and (213) and d_{123} in a HCP structure with $a = 2$ nm.

CHAPTER 3

INTERFERENCE AND DIFFRACTION

CONTENTS

3.1	Refraction and Reflection	94
3.2	Interference	96
3.3	Diffraction	105
Problems		112

3.1 REFRACTION AND REFLECTION

Refraction refers to the change in propagation direction of a wave due to a change in its medium. The phenomenon is mainly governed by the law of energy and momentum conservation. It is commonly observed when a wave passes from one medium to another at any angle other than 0° from the normal. Refraction of light is the most commonly observed phenomenon, but any type of wave can be refracted when it interacts with a medium. Refraction is also responsible for rainbows and for the splitting of white light into a rainbow-spectrum as it passes through a glass prism. A long object such as pencil or wood stick obliquely immersed in water looks bent due to refraction. This is because a light ray reflected from the tip of the object refracts as it leaves the surface of water. Thus, the ray conceived by our eyes looks as if it were reflected from a point other than the tip of the object. Understanding of this concept led to the invention of lenses and glasses.

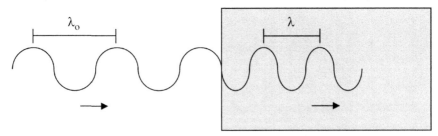

Refractive index, $n = \lambda_o/\lambda$

FIGURE 3.1 Definition of refractive index.

When a light is incident from free space into a matter, its speed is decreased. Since the frequency remains unchanged, its wavelength is also reduced inside the matter (Figure 3.1). The refractive index n of a substance is defined as

$$n = \frac{c}{\upsilon} = \frac{\lambda_o}{\lambda} \tag{3.1}$$

where c and λ_o are the speed and wavelength of light in vacuum, and υ and λ, the speed and wavelength of that light in the substance. Thus, vacuum has $n = 1$ as the reference medium. Air has a refractive index only slightly larger than unity. The refractive index of water is 1.33 at visible ranges, implying that light travels 1.33 times slowly in water than in vacuum or air. The refractive index also determines how light is bent, i.e., refracted, when entering a material. This is described by Snell's law of refraction: $n_1 \sin \theta_1 = n_2 \sin \theta_2$, where θ_1 and θ_2 are the angles of incidence and refraction of a ray crossing the interface between two media with refractive indices n_1 and n_2. In Figure 3.2, n_2 is assumed to be larger than n_1. When light is incident from a lower-index material into a higher-index material, the angle of refraction is smaller than the angle of incidence and the light is refracted toward the normal of the interface. When entering a medium with lower refractive index, the light is refracted away from the interface normal. The angle of reflection is the same as the angle of incidence. The elementary rules of refraction and reflection are deduced from the application of boundary conditions for electromagnetic waves. These rules can also be deduced from Fermat's principle; *the path taken between two points by a ray of light is the path that can be traversed in the least time.* In general, the incident wave is partially refracted and partially reflected. The refractive indices also determine how much of the light is reflected from the interface, and how much is refracted in a given situation. The details of this behavior are described by the Fresnel equations for refraction and reflection. For normal incidence, the reflectance is proportional to the square of the refractive index difference between two media.

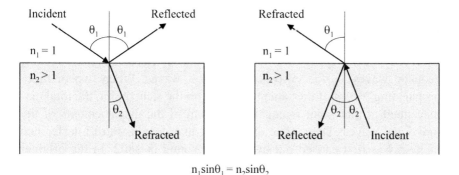

FIGURE 3.2 The angle of reflection is the same as the angle of incidence. The angle of refraction is related to the angle of incidence by the Snell's law: $n_1 \sin \theta_1 = n_2 \sin \theta_2$.

96   X-Ray Diffraction for Materials Research: From Fundamentals to Applications

The concept of refractive index is widely used within the full electromagnetic spectrum. The refractive index of a substance varies with the wavelength of incident electromagnetic radiation. All materials have a refractive index very close to 1 at X-ray wavelengths. As an example, water has $n = 1 - 2.6 \times 10^{-7}$ for X-ray radiation at $\lambda = 0.04$ nm. The refractive index lower than unity does not contradict the theory of relativity, which holds that no information-carrying signal can travel faster than the speed of light in vacuum. The refractive index is defined with respect to the phase velocity, which does not carry information. The phase velocity is not the same as the group velocity or the signal velocity. It means the speed at which the crests of the wave move and can be faster than the speed of light in vacuum, thereby giving a refractive index below 1. The analysis of materials using X-rays is much facilitated by the fact that the refractive index is nearly equal to 1 at X-ray regions for all materials. Since the incident beam is not refracted at the air-material interface, sample polishing is unnecessary. In addition, we do not have to separately calculate the propagation direction of X-ray beam within the material because it is basically the same as the incident direction. The dependence of refractive index on the frequency of electromagnetic wave is well described in many textbooks on optics and electrodynamics.[1-4,9,10]

3.2 INTERFERENCE

Interference is a phenomenon in which two waves superpose to form a resultant wave of higher or lower amplitude. Interference effects can be observed will all types of waves, for instance, electromagnetic waves, acoustic waves, surface waves, and matter waves. When two or more propagating waves of the same type meet on the same point, the total displacement at that point is equal to the sum of the displacements of the individual waves. The classic experiment that demonstrates interference of light was first carried out by Thomas Young in 1802. In the original experiment, sunlight was used as the source. Light was passed through a small pinhole so as to illuminate two narrow slits. An interference pattern of bright and dark fringes was observed in a screen placed behind the slits. Young was able to estimate the wavelength of different colors in the spectrum from the spacing of the fringes. His experiment played a major role in

Interference and Diffraction

the general acceptance of the wave theory of light. A key to the experiment is the use of a single pinhole to illuminate the slits. This provides the necessary mutual coherence between the lights that comes from the two slits.

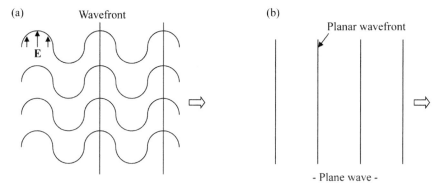

FIGURE 3.3 (a) Wavefront refers to the locus of points having the same phase. (b) The wavefronts of a plane wave are planes.

Wavefront means an imaginary surface joining all points of equal phase in a wave. The wavefronts of a plane wave are planes. Any wave propagates in the direction normal to its wavefront. A plane wave thus has a straight propagation direction, as shown in Figure 3.3. A small stone vertically dropped into a tranquil lake will generate a two-dimensional circular wave on the water surface, in which the wavefront is in the form of concentric circles. Similarly, an electromagnetic wave emitted radially from a point source can be considered as a spherical wave. The surface of a sphere becomes more flattened with increasing radius. Thus, a spherical wave will behave like a plane wave when it is far away from the source. As illustrated in Figure 3.3(a), the wavefronts are usually drawn along the crests of a wave, with the spacing between two adjacent wavefronts equal to one wavelength.

EXAMPLE 3.1

State how the wavefront changes when a plane wave is focused by a lens.

Answer: When a plane-wave beam is focused by a lens, the beam size decreases and then increases after being minimized at focus. Now, the wave has different propagation directions depending on the position within the beam. Since the wave propagates normal to its wavefront, a plane wave focused by a lens will exhibit the wavefront shapes shown in Figure 3.4.

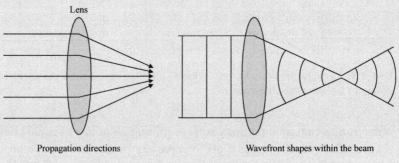

FIGURE 3.4 Wavefront shapes of a plane-wave beam focused by a lens.

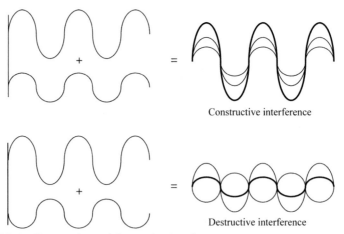

FIGURE 3.5 Constructive and destructive interference.

Because the frequency of electromagnetic waves is too high to be detected by currently available detectors, it is possible to observe only the intensity of an interference pattern. The intensity of an electromagnetic wave is proportional to its amplitude squared. Therefore, the superposition of two or more waves may yield a resultant intensity that is not simply the sum of the component intensities. A crucial factor determining the interference result is the difference in phase between the involved waves. Suppose that two harmonic waves with the same frequency are superposed. When the phase difference between these two waves is $\delta = 0, 2\pi, 4\pi, \ldots$, the resulting amplitude is maximized, whereas $\delta = \pi, 3\pi, 5\pi, \ldots$ yield a minimum amplitude (Figure 3.5). The former situation is referred to as *constructive interference*, in which the waves are in phase. In the latter *destructive interference*, the waves are 180° out of phase.

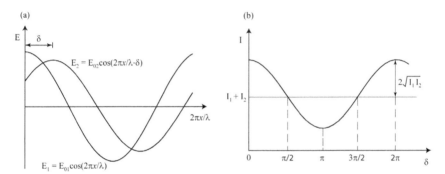

FIGURE 3.6 (a) Superposition of two harmonic waves E_1 and E_2. (b) Resultant intensity.

As a general case, we consider two electromagnetic waves E_1 and E_2 propagating along the positive x direction, with their electric fields directed along the y direction (Figure 3.6(a)). When these waves of the same frequency have amplitudes E_{01} and E_{02} and a phase difference of δ, the resultant disturbance is the linear superposition of the two waves and is expressed as

$$E = E_1 + E_2 = E_{01} \cos(2\pi x / \lambda) + E_{02} \cos(2\pi x / \lambda - \delta) \quad (3.2)$$

In dealing with the superposition of harmonic waves, the complex-number representation is mathematically simpler than the trigonometric manipulation. Keeping in mind that $e^{i\theta} = \cos\theta + i\sin\theta$, the relations of Eq. (3.2) can be alternatively described by the following complex notations.

100 X-Ray Diffraction for Materials Research: From Fundamentals to Applications

$$E_{com} = E_{01}e^{i2\pi x/\lambda} + E_{02}e^{i(2\pi x/\lambda - \delta)} = (E_{01} + E_{02}e^{-i\delta})e^{i2\pi x/\lambda} \quad (3.3)$$

Since E is simply the real part of E_{com}, we can take it after manipulation. The term in parentheses of Eq. (3.3) is also complex, and thus it can be expressed as follows.

$$E_{com} = (E_{01} + E_{02}e^{-i\delta})e^{i2\pi x/\lambda} = |E_{com}|e^{i\theta} e^{i2\pi x/\lambda} \quad (3.4)$$

where $|E_{com}|$ is the magnitude of E_{com}. Now we know that $E = |E_{com}|\cos(2\pi x / \lambda + \theta')$. θ' is a value depending on E_{01}, E_{02}, and δ. However, the intensity of the composite wave can be obtained without calculating this value. Multiplying each side of $(E_{01} + E_{02}e^{-i\delta}) = |E_{com}|e^{i\theta}$ by its complex conjugate leads to $|E_{com}|^2 = E_{01}^2 + E_{02}^2 + 2E_{01}E_{02}\cos\delta$. If we are concerned only with relative intensities, we can neglect the common proportionality constant and represent the component intensities as $I_1 = E_{01}^2$ and $I_2 = E_{02}^2$. The total intensity is then

$$I = I_1 + I_2 + 2\sqrt{I_1 I_2}\cos\delta \quad (3.5)$$

where $2\sqrt{I_1 I_2}\cos\delta$ is called *interference term*. The resultant intensity may be greater, equal to, or less than $I_1 + I_2$, depending on the value of the interference term, as plotted in Figure 3.6(b). The maximum intensity is obtained at $\cos\delta = 1$, when δ is an integer multiple of 2π. For $0 < \cos\delta < 1$, we have $I > I_1 + I_2$. The minimum intensity occurs when the waves are 180° out of phase, that is, when δ is an odd multiple of π. Interference refers to the interaction of waves that are coherent with each other. If the two waves are mutually incoherent, the phase difference δ varies with time in a random fashion. Then, the mean value of the interference term is zero, and there is no interference. The two waves must have the same polarization in order to maximize the interference effect. In particular, if the polarizations are mutually orthogonal, there is no interference again.

An interference fringe pattern is produced if two plane waves of the same frequency intersect at a non-zero angle θ. Interference is essentially an energy redistribution process. The energy lost at the destructive interference is regained at the constructive interference. The fringe pattern generated by two nonparallel plane waves is a series of straight lines. The fringes are observed wherever the two waves overlap. The fringe spacing increases with increasing wavelength and decreasing angle θ (see Example 3.2). A point source produces a spherical wave. If the light from two point sources overlaps, the interference pattern maps out the way in which the phase difference between the two waves varies in space. This depends on the wavelength and on the separation of the point sources. When the plane of observation is far away, the fringe pattern will be a series of nearly straight lines, since the waves will then be almost planar.

We have thus far been concerned with the interference between two beams. Multi-beam interference is more general. The most common method of producing a large number of coherent beams is by division of amplitude.

EXAMPLE 3.2

The superposition of two nonparallel plane waves of the same wavelength generates a one-dimensional interference pattern, as shown in Figure 3.7.

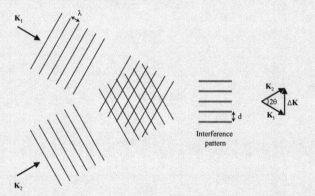

FIGURE 3.7 One-dimensional interference pattern generated by superposition of two nonparallel plane waves of the same wavelength.

This can be more easily understood with the concept of wave vector. The wave vector **K** is parallel to the propagation direction, i.e., normal to the wavefront, and has a magnitude of $2\pi/\lambda$. If two beams with the wave vectors $\mathbf{K}_1$ and $\mathbf{K}_2$ are superimposed, an interference pattern is generated with a vector $\Delta \mathbf{K}$ that is defined by the difference of the incident wave vectors, i.e., $\Delta \mathbf{K} = \mathbf{K}_2 - \mathbf{K}_1$. The magnitudes of these vectors are $K_1 = K_2 = 2\pi/\lambda$, $\Delta K = 2\pi/d$, where "d" is the period of the interference pattern. When the two beams have an intersection angle of 2θ, the magnitude of $\mathbf{K}_2 - \mathbf{K}_1$ is $4\pi \sin\theta/\lambda$. Then, we have

$$d = \frac{\lambda}{2\sin\theta} \qquad (3.6)$$

The period of the pattern, i.e., the spacing of the interference fringes is dependent on the wavelength and intersection angle of the interfering beams. For instance, when the angle 2θ between two interfering plane waves is 60°, the period of the resulting pattern is equal to the wavelength of the waves. Eq. (3.6) is equivalent to the Bragg law that governs the direction of X-ray diffraction, which will be discussed later throughout this book. In this example, we just emphasize the usefulness of the wave vector concept described above. Suppose that two He-Ne laser beams (λ = 632.8 nm in air) are symmetrically incident into a material of refractive index n at an external angle of 60° and generates an interference pattern inside the material, as shown in Figure 3.8. Then, find the period "d" by explaining how we can calculate it without knowing the refractive index of the material.

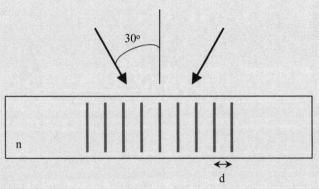

FIGURE 3.8 An interference pattern generated by two beams.

Answer: When a plane wave is incident into a material of refractive index n from free space, it is refracted toward the normal of the interface (Figure 3.9). As the angles of incidence and refraction are governed by the Snell's law: $\sin\theta_i = n\sin\theta_r$, the incident and refracted waves have different propagation directions. When the incident wave vector $\mathbf{K}_i$ has a magnitude of $2\pi/\lambda$, the magnitude of the refracted wave vector $\mathbf{K}_r$ becomes $2\pi n/\lambda$, since the wavelength inside the material is decreased to λ/n.

Interference and Diffraction

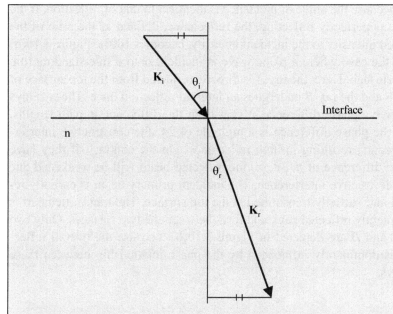

FIGURE 3.9 Wave vectors of the incident and refracted beams.

The Snell's law is rewritten as

$$\frac{2\pi}{\lambda}\sin\theta_i = \frac{2\pi n}{\lambda}\sin\theta_r \quad (3.7)$$

Both sides of Eq. (3.7) represent the components of the wave vectors tangential to the interface. In other words, the Snell's law states that the tangential components of the wave vectors are preserved. As the wave refracts, the direction and magnitude of its wave vector change. However, the tangential component of the wave vector remains unchanged. Therefore, the interference pattern formed inside a material has the same period as that obtained without the material. For $2\theta = 60°$, the period is equal to the wavelength, i.e., $d = 632.8$ nm.

The amplitude division occurs by multiple reflection between two parallel, partially reflecting surfaces. These surfaces might be semitransparent mirrors, or merely the two sides of a film or slab of transparent material. A plane wave reflected from a flat surface is also a plane wave. The rays *A* and *B* in Figure 3.10(a) maintain the same phase all the way along their paths because the angle of incidence is identical to that of reflection. If the surface is perfectly reflecting, the reflectance, defined as the ratio of the reflected intensity to the incident intensity, becomes 100%. Figure 3.10(b) shows the case where a plane wave is incident onto a free-standing thin dielectric slab. Here, the ray A is directly reflected from the top surface of the slab and the ray B undergoes an internal reflection once. The two rays may have a phase difference arising from the difference in path lengths. When the phase difference is a multiple of 2π, the constructive interference occurs, resulting in high reflectance. On the contrary, if they have a phase difference of π, 3π, ... the reflected beam will be weakened due to the destructive interference. The incident primary beam is partially reflected and partially transmitted at the top surface. The transmitted part is subsequently reflected back and forth between the two surfaces. Only two rays A and B are depicted in Figure 3.10(b), because the overall reflectance is dominantly influenced by the phase relationship between these two rays.

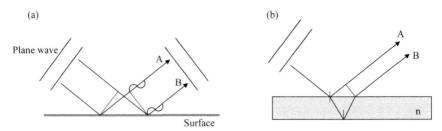

FIGURE 3.10 (a) A plane wave reflected from a flat surface is also a plane wave, since the rays **A** and **B** are always in phase. (b) Reflection from a thin slab.

Once two media have different refractive indices, reflection from their interface is inevitable. Air is also a medium of $n = 1$. For normal incidence, the reflectance of a material is given by $R = (n-1)^2 / (n+1)^2$. For example, a glass ($n = 1.5$) in air has $R = \sim 4\%$. A typical method of suppressing the reflectance of a bulk material is to coat it with a thin film of different refractive index so that the component waves reflected from the air-film and film-material interfaces destructively interfere (Figure 3.11). Interference is very common in our daily life. For example, the colors seen in a soap bubble arise from interference of light reflecting off the front and back surfaces of the thin soap film. Depending on the thickness of the film, different colors interfere constructively and destructively.

Interference and Diffraction

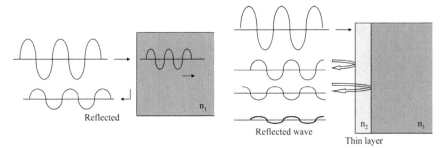

FIGURE 3.11   Principle of anti-reflection coating.

3.3   DIFFRACTION

There is no way to satisfactorily explain the difference between *interference* and *diffraction*. Although not always appropriate, it is sometimes customary to speak of interference when treating the superposition of a few waves and diffraction when considering a large number of waves. The propagation of light in media is often explained by **Huygens**'s principle; *every point on a primary wavefront serves as the source of secondary spherical waves (or wavelets) and these secondary waves constitute the new wavefront*. The essential features of diffraction can be qualitatively described by Huygens's principle either. A simple illustration of the principle is shown in Figure 3.12 for a plane wave.

This principle states that the propagation of a light wave can be predicted by assuming that each point of the wavefront acts as the source of a secondary wave spreading out in all directions. Therefore, we can suppose that a planar wavefront contains a lot of (imaginary) oscillators and each oscillator generates a spherical wave. Since every point on a wavefront serves as the source, the oscillators are considered very closely spaced. Thus, the new wavefront formed by these secondary waves will also be planar. A plane wave propagates through a medium in this way. According to Huygens's principle, there would have been a backward wave moving toward the source, which is not observed in reality. When a light wave propagates inside a material, each atom of the material interacts with an incident primary wavefront. Therefore, the atoms can be regarded as a point source of the scattered secondary wavelets. However, things are not quite clear when the principle is applied to the propagation of light through a vacuum. Nevertheless, the imaginary oscillator model proposed by Huygens fits in well with many optical phenomena.

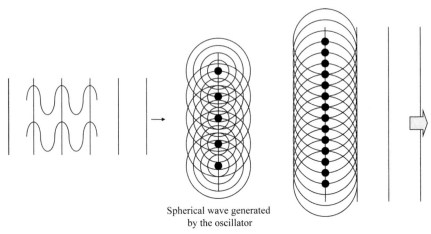

FIGURE 3.12 Propagation of a plane wave via Huygens's principle.

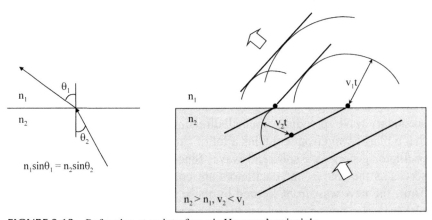

FIGURE 3.13 Refraction at an interface via Huygens's principle.

As an example, Figure 3.13 illustrates the application of Huygens's principle to the law of refraction, where a plane wave propagates from a medium with higher index (n_2) to a medium with lower index (n_1). The incident wave generates secondary wavelets and these wavelets overlap to form the new wavefront. Since $n_2 > n_1$, the lower-index medium has a higher light speed, i.e., $v_2 > v_1$. This means that in a given time t, the wavelet travels a longer distance in the lower-index medium than in the higher-index medium. Therefore, the wavefront formed in the lower-index medium is more inclined toward the interface normal, making the angle of refraction larger than the angle of incidence. It can be easily proved that the result is consistent with the Snell's law of refraction. When a beam is focused by a lens, the beam size at focus is decreased as the focal length of the lens decreases. Since the beam shape is symmetric with respect to the focal point (refer to Figure 3.4), a more tightly focused beam is more widely spread. This can also be explained with the oscillator model. If the areal density of oscillators is fixed constant, a small beam containing only a few oscillators will be highly divergent. As the beam becomes bigger, the number of oscillators increases as much. Thus, more secondary waves are generated and this makes the wavefront more planar (Figure 3.14).

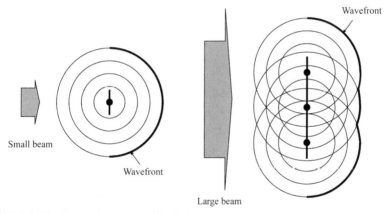

FIGURE 3.14 Beam size vs. wavefront shape.

A classic example of diffraction is the spreading out of a wave passing through a small aperture, as depicted in Figure 3.15(a). There are numerous point sources on the incoming plane wave. However, secondary waves only from some points in front of the aperture can pass through it and those from the others are blocked, as depicted in Figure 3.15(b). As the aperture gets smaller, the number of unobstructed points decreases. This makes an incoming wave more widely spread. When the aperture is a long, narrow slit, a cylindrical wave will emerge, as illustrated in Figure 3.16(a). Suppose that a plane wave falls on two narrow, parallel, and closely spaced slits. Then, the slits can serve as secondary sources. Two cylindrical waves coming from the slits are always in phase along certain directions and 180° out of phase along some others. Thus, if a screen is placed far away from the slits, alternating bright and dark interference fringes will be produced on the screen (Figure 3.16(b)). Here we see again that there is no physical difference between interference and diffraction. When a light beam is confronted with a periodic structure, it is split into several beams traveling in specific directions. This behavior is also called diffraction, in which the periodic structure plays a role of diffraction grating. Any periodic structures (e.g., periodically arranged apertures, surface relief pattern, and refractive-index variation) can serve as an effective diffraction grating, once the grating period is similar to the wavelength of an electromagnetic wave in question.

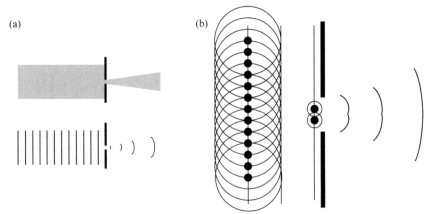

FIGURE 3.15 (a) A classic example of diffraction: the spreading out of a wave passing through a small aperture. (b) Secondary waves only from some points in front of the aperture can pass through it and those from the others are blocked.

Interference and Diffraction

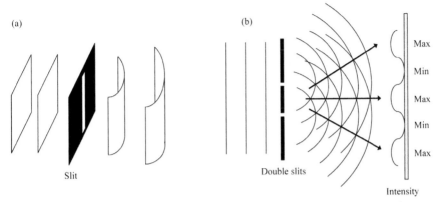

FIGURE 3.16 (a) Cylindrical wave emerging from a long, narrow slit. (b) Interference between two cylindrical waves emerging from the double slits.

As stated earlier, all materials have a refractive index $n = 1$ at X-ray wavelengths. Thus, it is impossible to generate a refractive-index grating inside or on the surface of a material for X-rays. Since crystalline materials comprise periodically arranged atoms, they already have embedded-diffraction gratings for X-rays. The origin of X-ray diffraction is scattering by these atoms. The atoms in a crystal scatter incident X-rays in all directions but the amplitude of the scattered wave depends on the scattering direction (Figure 3.17). The wave scattered by a single atom has an extremely weak electric field compared to that of the incident X-ray beam. However, when the scattered waves from a number of atoms are in phase and constructively interfere, the scattering intensity becomes considerably high.

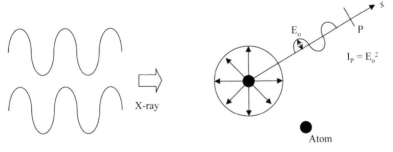

FIGURE 3.17 Scattering of X-rays by an atom.

As a simple bridge between the interference and diffraction, consider a linear array of N scattering centers illustrated in Figure 3.18. We are concerned with the scattering along the x-direction and measure its intensity at some distant point P. If the spatial extent of the array is comparatively small, the amplitudes of the waves arriving at P will be essentially equal. Then, the total electric field at P may be written in its complex exponential form as follows.

$$E = E_o e^{i(2\pi x/\lambda)} + E_o e^{i(2\pi x/\lambda - \delta)} E_o e^{i(2\pi x/\lambda - 2\delta)} + \ldots + E_o e^{i(2\pi x/\lambda - (N-1)\delta)} \quad (3.8)$$

where E_0 is the amplitude of the scattered wave and δ is the phase difference between the waves scattered from two adjacent centers. Eq. (3.8) can be rewritten as

$$E = E_o e^{i(2\pi x/\lambda)}(1 + e^{-i\delta} + e^{-i2\delta} + \ldots + e^{-i(N-1)\delta}) \quad (3.9)$$

The parenthesized geometric series has the value of $(1 - e^{-iN\delta})/(1 - e^{-i\delta})$, which can be rearranged into the form $[e^{-iN\delta/2}(e^{iN\delta/2} - e^{-iN\delta/2})]/[e^{-i\delta/2}(e^{i\delta/2} - e^{-i\delta/2})]$ or equivalently

$$e^{-i(N-1)\delta/2} \frac{\sin(N\delta/2)}{\sin(\delta/2)} \quad (3.10)$$

The total field then becomes

$$E = E_o e^{i(2\pi x/\lambda)} e^{-i(N-1)\delta/2} \frac{\sin(N\delta/2)}{\sin(\delta/2)} = E_o \frac{\sin(N\delta/2)}{\sin(\delta/2)} e^{i[2\pi x/\lambda - (N-1)\delta/2]} \quad (3.11)$$

The total intensity is

$$I = E_o^2 \frac{\sin^2(N\delta/2)}{\sin^2(\delta/2)} \quad (3.12)$$

Interference and Diffraction

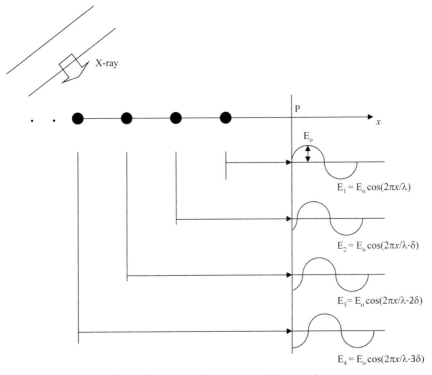

FIGURE 3.18 Scattering of X-rays by a linear array of N scattering centers.

As shown in Figure 3.19(a), the intensity depends strongly on the phase difference δ. It rises to maxima $I = N^2 E_o^2$ at $\delta = m2\pi$, where m is an integer. The intensity is essentially zero except for the limited range of $m2\pi - 2\pi/N < \delta < m2\pi + 2\pi/N$. As the number of scattering centers, N, increases, the peaks become stronger and narrower. This means that with more scattering centers, the peak intensity increases but the condition for constructive interference gets stricter. Figure 3.19(b) compares the peak shapes for $N = 10$ and $N = 20$. For very large N, a small deviation from $\delta = m2\pi$ will make the intensity drop to zero.

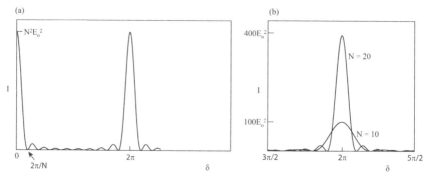

FIGURE 3.19 (a) Dependence of intensity on the phase difference δ. (b) Peak shapes for two different numbers of scattering centers.

PROBLEMS

3.1. It has been shown in Example 3.2 that the superposition of two nonparallel plane waves generates a one-dimensional interference pattern and the pattern period can be derived by considering their wave vectors. By the same token, we can generate a two-dimensional pattern using three plane waves. In Figure 3.20, three plane-wave beams at $\lambda = 500$ nm are incident onto a photosensitive thin film with the following wave vectors.

$$\mathbf{K}_1 = \frac{2\pi}{\lambda}(\frac{1}{2}\mathbf{i} + \frac{\sqrt{3}}{6}\mathbf{j} - \frac{2}{\sqrt{6}}\mathbf{k})$$

$$\mathbf{K}_2 = \frac{2\pi}{\lambda}(-\frac{1}{2}\mathbf{i} + \frac{\sqrt{3}}{6}\mathbf{j} - \frac{2}{\sqrt{6}}\mathbf{k})$$

$$\mathbf{K}_3 = \frac{2\pi}{\lambda}(-\frac{2\sqrt{3}}{6}\mathbf{j} - \frac{2}{\sqrt{6}}\mathbf{k})$$

where $\mathbf{i}$, $\mathbf{j}$, and $\mathbf{k}$ are unit vectors along the x, y, and z directions. Then, what are the shape and period of a two-dimensional pattern generated in the film? Refer to the Refs. [31–35] for more details on the multi-beam interference and applications.

Interference and Diffraction 113

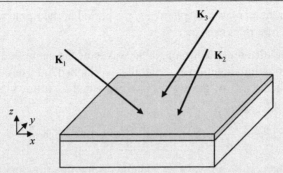

FIGURE 3.20 Three-beam interference.

3.2. The following graph shows a ripple structure observed in the transmission spectrum of a water layer formed on slide glass. The bottoms in the ripple were found at λ = 497 nm, 520 nm, 545 nm, 574 nm, 606 nm, 641 nm, 681 nm, 722 nm, etc. Water has n = 1.33 and glass, n = 1.50.

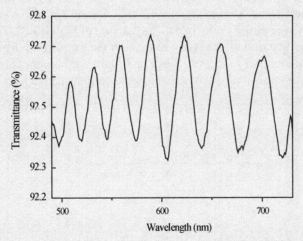

FIGURE 3.21 Transmission spectrum of a water layer on slide glass.

(a) Estimate the thickness of this water layer from the given data.

(b) In a thin film/substrate structure, the height of the ripple (i.e., the transmittance difference between the top and bottom of the ripple) increases as the refractive index difference between the film and the substrate increases. Explain how the period of the ripple will change if the film thickness increases.

3.3. Three different electromagnetic waves of the same frequency and polarization direction are spatially overlapped, interfering with one another. Each of the waves is represented as follows at a certain instant.

$$E_1 = E_o \cos\left(\frac{2\pi x}{\lambda}\right), E_2 = 2E_o \cos\left(\frac{2\pi x}{\lambda} - \frac{\pi}{6}\right), E_3 = E_o \cos\left(\frac{2\pi x}{\lambda} - \frac{\pi}{4}\right)$$

Express the combined wave as a single cosine function and find its intensity in terms of E_o.

3.4. Diffraction, caused by the wave nature of light, can be explained with the oscillator model. Suppose that the distance between oscillators is similar to the wavelength. Then, which one between visible and infrared lights will be spread more widely when they are passing through a small aperture of fixed size?

3.5. According to Fermat's principle, a light beam traversing an interface does not take a straight line but travels along a path that takes the least time. Derive the law of refraction of $n_1 \sin \theta_1 = n_2 \sin \theta_2$ by applying Fermat's principle to the diagram of Figure 3.22. To solve the problem, it is necessary to express the transit time from A to B with respect to the variable x and find its minimum value. The smallest transit time will then coincide with the actual path.

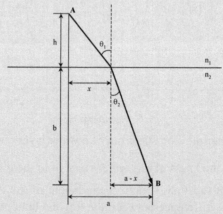

FIGURE 3.22 Fermat's principle applied to refraction.

PART II
THEORY OF X-RAY DIFFRACTION

CHAPTER 4

DIRECTIONS OF X-RAY DIFFRACTION

CONTENTS

4.1	Introduction	118
4.2	Bragg Law	120
4.3	Laue Equations	127
4.4	Diffraction Condition in Reciprocal Space	130
4.5	Off-Bragg Angle Diffraction	134
4.6	Electron Diffraction	141
Problems		144

4.1 INTRODUCTION

In the previous chapter, we mentioned that X-ray diffraction is essentially a scattering phenomenon in which a large number of atoms are involved. Since the atoms in a crystal are periodically arranged, the X-rays scattered by these atoms can be in phase and constructively interfere in a few directions. If the atoms were not arranged in a regular, periodic fashion, the rays scattered by them would have a random phase relationship to one another. Under this condition, neither constructive nor destructive interference takes place, and the scattering intensity in a particular direction will be simply the sum of the intensities of all the rays scattered in that direction. The intensity of electromagnetic radiation is proportional to the square of its amplitude. If there are N scattered rays, each of amplitude E_0, the total amplitude is NE_0 when they are all in phase. Then, the intensity of the scattered beam becomes $N^2 E_0^2$. On the contrary, when the scattered rays have a random phase relationship, the scattering intensity is NE_0^2, which is N times smaller than the former case. Since solid materials contain $10^{22}-10^{23}$ atoms/cm^3, N is a very large number in X-ray diffraction.

The diffraction of light had been well understood before X-rays were discovered. It had already been known that a visible light is diffracted whenever it encounters regularly spaced obstacles, apertures, or engraved structures having a period of the same order of magnitude as the wavelength. The German physicist Max von Laue was the first to use X-rays to study the arrangement of atoms in crystals. The precise nature of X-radiation, discovered by W. C. Röntgen in 1895, had not yet been determined when von Laue initiated his study of X-rays in 1912. He guessed that if crystals were composed of regularly arranged atoms, and if X-rays were electromagnetic waves of wavelength similar to the distance between the atoms, it would be possible to diffract X-rays by crystals. Experiments were carried out to prove this hypothesis and his co-workers succeeded in recording diffraction spots from a copper sulfate crystal on a photographic plate. These experiments simultaneously confirmed the wave nature of X-rays and the periodic arrangement of atoms within a crystal and also made it possible to measure the wavelength of X-rays with great accuracy. The significance of these experiments was immediately recognized by the scientific community. The pioneering work by Laue and his colleagues gave scientists a new tool for investigating the atomic structure of matter and established the so-called "*X-ray crystallography*" area. It was X-ray diffraction study that enabled the molecular structures of DNA and RNA to be revealed in 1950s.

In the same year of 1912, the British physicist William L. Bragg and his father William H. Bragg expressed the condition for X-ray diffraction in a much simpler mathematical form than that explained by von Laue. They found that some crystals, at certain wavelengths and incident angles, produced intense peaks of reflected X-rays known as *Bragg peaks*. W. L. Bragg explained this result by modeling the crystal as a set of discrete parallel planes separated by a constant parameter. It was proposed that the incident X-rays would produce a Bragg peak if their reflections off the various planes interfered constructively. The interference is constructive when the phase difference is a multiple of 2π; this condition can be expressed by Bragg's law. In the following year, W. L. Bragg revealed the structures of NaCl, KCl, KBr, and KI, which all have the NaCl structure. This was the first to completely analyze the structure of crystalline material. W. L. Bragg and W. H. Bragg were awarded the Nobel Prize in physics in 1915 for their pioneering work in determining crystal structures. They are the only father-son team to jointly win the prize. W. L. Bragg was then 25 years old, the youngest Nobel laureate. Although it is very simple, Bragg's law confirmed the existence of real particles at the atomic scale, as well as providing a powerful tool for examining crystals. The early work by Bragg had an immeasurable impact on the development of X-ray diffraction and modern diffractive optics. It is not surprising that Eq. (3.6), which dictates the period of two-beam interference pattern, is the same as the Bragg law for X ray diffraction.

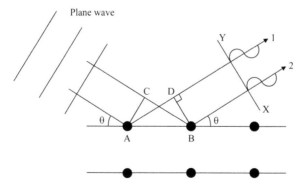

FIGURE 4.1 Rays 1 and 2 are always in phase when the incident and scattering angles are identical with respect to the atomic planes.

4.2 BRAGG LAW

Let's suppose that an X-ray beam of plane wave is incident into a crystal with atoms arranged on a set of parallel planes and the incident beam makes an angle of θ with the crystal plane, as shown in Figure 4.1. While the X-ray beam is scattered in all directions, we here consider the case where scattering occurs at an angle of θ with respect to the plane. The ray **1** scattered by the atom located at point A always has the same phase as the ray **2** scattered by the atom at B because the length AD is equal to the length CB. In fact, there is no path length difference between the rays scattered by the atoms of a plane when the angle of scattering equals the angle of incidence. The scattered rays **1** and **2** of Figure 4.1 are spatially apart from each other. The diffraction intensity is determined by the phase relationship between the spatially overlapped rays. In Figure 4.2, the ray **1S**, scattered from the first plane, spatially overlaps the ray **3S** scattered from the second plane. The diffraction intensity is ultimately decided by the phase relationship between these rays. The rays **3S** and **2S**, both scattered from the second plane, have no path length difference. Therefore, the phase difference between the rays **1S** and **3S** is identical to that between the rays **1S** and **2S**. It is easier to calculate the phase difference with the rays **1S** and **2S**, rather than the rays **1S** and **3S**.

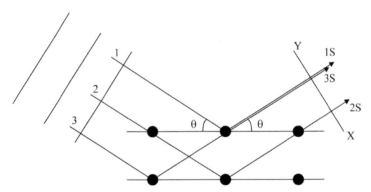

FIGURE 4.2 The path length difference between rays 1S and 3S is identical to that between rays 1S and 2S.

Directions of X-ray Diffraction 121

Figure 4.3(a) depicts how the condition for diffraction is derived. The difference in path length between the rays 1S and 2S is $CA + AD = 2d\sin\theta$, where d is the interplanar spacing. When this path length difference is equal to an integer multiple of the wavelength, the two rays are completely in phase and an arbitrary plane XY normal to the scattering direction becomes a planar wavefront. The condition for diffraction is then

$$n\lambda = 2d\sin\theta \qquad (4.1)$$

where n is an integer. This relation was first derived by W. L. Bragg and is known as the Bragg law or equation. It states that the incident and diffracted beams are coplanar with the normal to the lattice planes and equally inclined at $90°-\theta$ to it and that the angle θ (called the Bragg angle) is related to the X-ray wavelength and to the interplanar spacing. A number of planes are involved in scattering because the X-ray beam size is much larger than the interatomic distance. An alternative way of looking at the Bragg equation is that the diffracted beam can be regarded as a reflection of the incident beam by a set of lattice planes. Bragg considered first how the X-rays scattered by all the lattice points (or atoms) in a single plane might be in phase. The condition for optical reflection is that the angle of incidence is equal to the angle of reflection (see Figure 3.10(a)). This ensures that the waves scattered by all points in that plane are in phase with one another, as illustrated in Figure 4.2. In general, the waves reflected from successive lattice planes will not be in phase. Figure 4.3(b) shows reflection from two adjacent planes. The waves reflected from the upper plane have a shorter path length than those reflected from the lower plane. When the path length difference is equal to a whole number of wavelengths, the waves reflected from successive planes are in phase and reinforce one another. In other words, the X-rays scattered by all the atoms in all the planes constructively interfere to form a diffracted beam in the given direction. The waves scattered in other directions are out of phase and cancel out with one another. In both of the optical reflection and Bragg reflection (i.e., X-ray diffraction), the angle of incidence is equal to the angle of reflection. However, two phenomena are fundamentally different. The optical reflection occurs in a very thin surface layer. On the contrary, the diffracted beam from a crystal is built up of waves scattered by all the atoms of the crystal irradiated by the incident X-ray beam. While the reflection of visible light can take place at any angles of incidence, the diffraction of monochromatic X-rays is possible only at specific angles of incidence that satisfy the Bragg law.

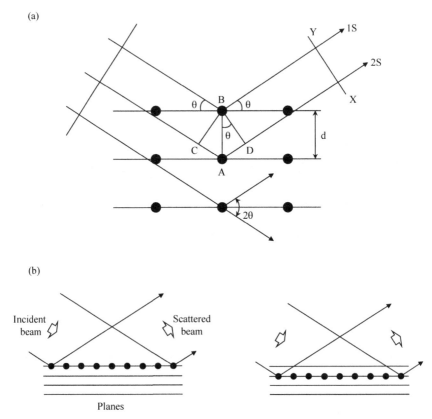

FIGURE 4.3 (a) Diffraction of X-rays by a crystal. (b) Reflection from successive lattice planes.

The integer n of Eq. (4.1), which is called the order of reflection, can take on any value unless $\sin\theta$ exceeds unity. Therefore, for fixed λ and d, there may be several angles of incidence $\theta_1, \theta_2...$, which correspond to $n = 1, 2, ...$ For the first-order reflection ($n = 1$), the scattered rays **1S** and **2S** of Figure 4.3(a) would differ in path length by λ. For $n = 2$, their path length difference would be twice, i.e., 2λ. Consider the first-order and second-order reflections from (001) planes, as shown in Figure 4.4. If the angles of incidence for the two reflections are θ_1 and θ_2, respectively, we obtain the following relations.

$$\lambda = 2d_{001} \sin \theta_1$$

$$2\lambda = 2d_{001} \sin \theta_2 \qquad (4.2)$$

The second-order reflection of Eq. (4.2) can be alternatively expressed as $\lambda = 2(d_{001}/2)\sin\theta_2 = 2d_{002}\sin\theta_2$. This means that the second-order reflection from the (001) planes is equivalent to the first-order reflection from (002) planes. In Figure 4.4, the dotted plane midway between the (001) planes corresponds to part of the (002) set of planes. An nth-order reflection from (hkl) planes may be regarded as a first-order reflection from the ($nh\ nk\ nl$) planes with $1/n$th the spacing of the former. It is conventional to represent the reflection from the (hkl) planes by hkl without parentheses. Although 002, 003, and 004 reflections are equivalent to the second-, third-, and fourth-order reflections from the (001) planes, it is more general to view them as the first-order reflections from the (002), (003), and (004) planes. This allows us to write the Bragg law of Eq. (4.1) simply as

$$\lambda = 2d \sin \theta \qquad (4.3)$$

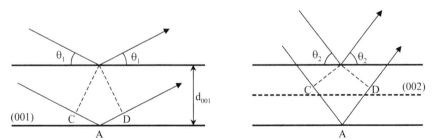

FIGURE 4.4 First-order and second-order reflections from (001) planes. The second-order 001 reflection is equivalent to the first-order 002 reflection.

This new form of Bragg law is more dominantly used in X-ray diffraction. The Bragg law can be experimentally applied in two different ways. By using monochromatic X-rays of known wavelength λ and measuring θ, we can determine the spacing d of various planes. Alternatively, a crystal with planes of known spacing d can be used to determine the radiation wavelength λ. Most applications of X-ray diffraction are associated with the measurement of the diffraction angle (2θ rather than θ is experimentally measured). The essential components of typical X-ray diffractometer are shown in Figure 4.5. A collimated beam from the X-ray source is incident onto a sample stationed on the sample holder, which may be set at any desired angle to the incident beam. A detector measures the intensity of the diffracted beam; it can be rotated around the sample and set at any desired angular position. The sample holder can also be rotated around its center independently of or in conjunction with the detector. The diffractometer measures the angle 2θ between the incident and detected beams. In the symmetric scan (often called θ-2θ scan or 2θ-θ scan), the angle θ between the incident beam and the sample holder is maintained at half the measured diffraction angle 2θ, as depicted in Figure 4.5.

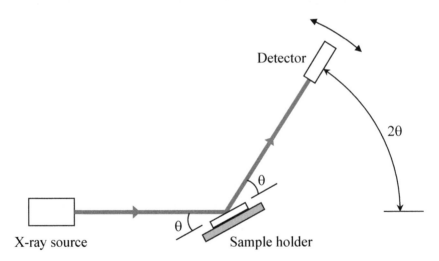

FIGURE 4.5 Essential components of typical diffractometer.

EXAMPLE 4.1

A plate sample was cut from a single crystal of simple cubic structure (a = 3 Å) so that its surface is parallel to (001) planes. When this sample is θ-2θ scanned in the range of 2θ = 20–120° using a Cu K_α line (λ = 1.54 Å) as the X-ray source, what will the obtained diffraction pattern look like?

Answer: The surface is parallel to the (001), (002), (003), etc., planes whose interplanar distances are d_{001} = 3 Å, d_{002} = 1.5 Å, d_{003} = 1 Å, etc. Since sinθ < 1, the basic condition for diffraction at any angle is $d > \lambda/2$. However, d_{004} = 0.75 Å < $\lambda/2$. Thus, 004 and higher-order reflections are impossible. We will have three diffraction peaks denoted as 001, 002, and 003 reflections (Figure 4.6). Their positions (i.e., diffraction angles) are easily obtainable from Eq. (4.3); $2\theta_{001}$ = 29.74°, $2\theta_{002}$ = 61.76°, $2\theta_{003}$ = 101.06°.

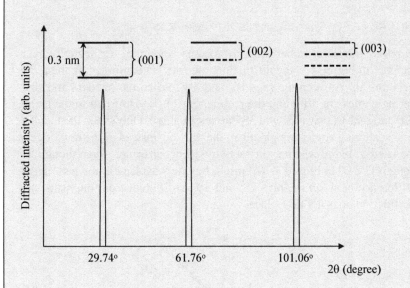

FIGURE 4.6 Expected diffraction pattern.

EXAMPLE 4.2

In the Bragg law of $\lambda = 2d\sin\theta$ for X-ray diffraction, θ refers to the incident and scattering angles with respect to the reflecting planes. Thus, diffraction can occur only at the same angle as the incident angle, as depicted in Figure 4.7(a) where the path length $CA + AD$ equals the wavelength λ. Consider a configuration shown in Figure 4.7(b). Even when the scattering angle θ_2 is different from the incident angle θ_1, the path length difference between the rays 1S' and 2S', i.e., the length $C'A + AD'$, may be equal to one wavelength. In this case, however, diffraction does not take place in the θ_2 direction. Explain why?

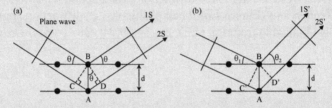

FIGURE 4.7 (a) Symmetric and (b) nonsymmetric scatterings.

Answer: Interference between rays occurs when they are spatially overlapped. In Figure 4.8, the diffraction intensity is determined by the phase relationship between the rays 1S' and 3S'. When the incident and scattering angles are different, the path length FAD' is unequal to the length EB and thus the rays 2S' and 3S' are not in phase. Only when the incident and scattering angles are identical, the ray 2S', instead of the ray 3S', can be used to judge whether diffraction will occur or not. Even though the length $C'AD'$ in Figure 4.7(b) might be one wavelength, the path length difference between the rays 1S' and 3S' is different from this value and diffraction does not take place.

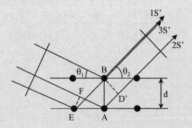

FIGURE 4.8 Nonsymmetric scattering; rays 2S' and 3S' are not in phase.

4.3 LAUE EQUATIONS

Although the Bragg law well describes the necessary condition for diffraction in a very simple manner, we take a brief look at Laue equations as well. Consider a row of scattering centers separated by a repeat distance "a", as shown in Figure 4.9. Let a monochromatic X-ray beam of wavelength λ is incident on the row at an angle α_o and scattered at an angle α. The difference in path length between rays scattered at point A and D is $AB - CD$. If the incident rays are originally in phase, this path length difference should be some integral multiple of the wavelength for constructive interference to occur. This leads to the following relation:

$$(AB - CD) = a(\cos\alpha - \cos\alpha_o) = h'\lambda \tag{4.4}$$

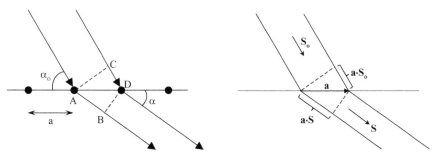

FIGURE 4.9 A row of scattering centers separated by a repeat distance a. The incident and scattered beams are represented by unit vectors S_o and S, respectively, and the repeat distance, by a unit cell vector $\mathbf{a}$.

where h' is an integer. This relation can be rewritten in vector forms. If we replace the repeat distance "a" with a unit cell vector $\mathbf{a}$ and let the incident and scattered beams be represented by unit vectors S_o and S, respectively, then $a\cos\alpha = \mathbf{a}\cdot\mathbf{S}$ and $a\cos\alpha_o = \mathbf{a}\cdot\mathbf{S}_o$. The above Eq. (4.4) is thus rewritten as

$$\mathbf{a}\cdot(\mathbf{S} - \mathbf{S}_o) = h'\lambda \tag{4.5}$$

For a three-dimensional crystal with the unit cell vectors $\mathbf{a}$, $\mathbf{b}$, and $\mathbf{c}$, we have two more Laue equations

$$\mathbf{b} \cdot (\mathbf{S} - \mathbf{S}_o) = k'\lambda$$
$$\mathbf{c} \cdot (\mathbf{S} - \mathbf{S}_o) = l'\lambda \qquad (4.6)$$

where k' and l' are also integers. The three Laue equations of Eq. (4.5) and (4.6) should be simultaneously satisfied for diffraction to occur. This condition can be met only when

$$\frac{\mathbf{S} - \mathbf{S}_o}{\lambda} = h'\mathbf{a}^* + k'\mathbf{b}^* + l'\mathbf{c}^* \qquad (4.7)$$

The dot products of Eq. (4.7) with $\mathbf{a}$, $\mathbf{b}$, and $\mathbf{c}$ will lead to Eq. (4.5) and (4.6). The three Laue equations give the necessary conditions for an incident beam to be diffracted by a crystal.

The Bragg law is given in Eq. (4.3) in a simple scalar form. The Bragg equation for (hkl) planes can be reformatted as

$$\frac{2\sin\theta_B}{\lambda} = \frac{1}{d_{hkl}} \qquad (4.8)$$

Figure 4.10 shows a graphical configuration for the Bragg law in which the incident and diffracted beams are represented by the vector $\mathbf{S}/\lambda$ and $\mathbf{S}_o/\lambda$, respectively. Here, $\mathbf{S}_o$ and $\mathbf{S}$ are unit vectors along the given directions. Then, the magnitude of $(\mathbf{S} - \mathbf{S}_o)/\lambda$ is equal to the left-hand side of Eq. (4.8) and the magnitude of $\mathbf{H}_{hkl}$ is $1/d_{hkl}$. $(\mathbf{S} - \mathbf{S}_o)/\lambda$ and $\mathbf{H}_{hkl}$ have the same directions because both are perpendicular to the reflecting planes. The Bragg law in vector form is then given by

$$\frac{\mathbf{S} - \mathbf{S}_o}{\lambda} = \mathbf{H}_{hkl} \qquad (4.9)$$

It can be seen that Eq. (4.7) is simply the Bragg law for the plane ($h'k'l'$). The integers $h'k'l'$ of the Laue equations are the Miller indices of the corresponding reflection. The Bragg equation provides a simple and convenient statement of the geometry of the diffraction of X-rays by crystals. It is the fundamental equation of X-ray crystallography. Its application in a variety of situations will be explored in the later sections of this chapter and in the following chapters.

Directions of X-ray Diffraction

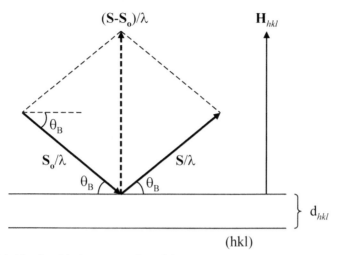

FIGURE 4.10 Graphical representation of the Bragg law.

It is worthy to note the analogy between Eq. (4.9) and the vector relation $\Delta \mathbf{K} = \mathbf{K}_1 - \mathbf{K}_2$ for two-beam interference, which was described in Example 3.2. Two plane-wave beams represented by the wave vectors $\mathbf{K}_1$ and $\mathbf{K}_2$ generate a one-dimensional interference pattern characterized by the vector $\Delta \mathbf{K}$. The magnitudes of these vectors are $K_1 = K_2 = 2\pi/\lambda$, $\Delta K = 2\pi/d$, where "d" is the period of the interference fringes. The two relations of $\Delta \mathbf{K} = \mathbf{K}_1 - \mathbf{K}_2$ and Eq. (4.9) are basically the same. The common factor "2π" may be or may not be included in the expressions. It is usually included in optics and solid state physics, but not in X-ray crystallography. Just as $\mathbf{H}_{hkl}$ is normal to the reflecting planes, $\Delta \mathbf{K}$ is also perpendicular to the generated interference fringes. The difference between two-beam interference and X-ray diffraction is that the former concerns the formation of a new grating with two incident beams, while the latter is to generate a secondary beam through the interaction between an X-ray beam and an already existing grating (a set of atomic planes). The equations describing two phenomena are identical. Two X-ray beams can generate a one-dimensional interference pattern. Of course, a visible light can be diffracted from a periodic structure once the Bragg law is satisfied.

4.4 DIFFRACTION CONDITION IN RECIPROCAL SPACE

The reciprocal lattice can provide a simple graphical representation of the Bragg law, as illustrated in Figure 4.11. The incident beam vector $\mathbf{S}_o/\lambda$ is first drawn to the origin of the reciprocal lattice. A sphere of radius $1/\lambda$ is then drawn centered on the initial point of the incident beam vector. This sphere is known as the sphere of reflection or *Ewald sphere*. The condition for diffraction is satisfied when any reciprocal lattice point *hkl* falls on the surface of this sphere. The direction of the diffracted beam is given by the vector $\mathbf{S}/\lambda$ drawn from the origin of the sphere to the point *hkl*. This construction is known as the Ewald construction. It is evident that the relation of the three vectors $\mathbf{S}_o/\lambda$, $\mathbf{S}/\lambda$, and $\mathbf{H}_{hkl}$ is that of the Bragg law given by Eq. (4.9). Although Figure 4.11 illustrates this relation schematically in two dimensions, the Ewald construction is valid in three dimensions. The geometric meaning of Figure 4.11 is that if the origin of the reciprocal lattice is placed at the tip of $\mathbf{S}_o/\lambda$, then diffraction will occur only for the reciprocal lattice points that lie on the surface of the Ewald sphere. If the incident beam is a white beam, with a wavelength range $\lambda_{min} \leq \lambda \leq \lambda_{max}$, there will be a nest of Ewald spheres of radii $1/\lambda_{max} \leq 1/\lambda \leq 1/\lambda_{min}$.

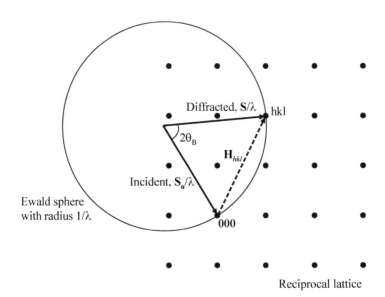

FIGURE 4.11 Diffraction condition in reciprocal space.

Directions of X-ray Diffraction

The diffraction condition in reciprocal space may be better understood with some examples. When the reflecting (*hkl*) plane and the X-ray wavelength are fixed (Figure 4.12), the beam should be incident at a specific angle in order for diffraction to occur. For other angles of incidence, the $\mathbf{H}_{hkl}$ vector cannot terminate on the surface of the Ewald sphere. Suppose that white X-rays are incident into the fixed (*hkl*) planes at a specific angle. In this case, the radius of the Ewald sphere is not fixed but variant. Thus, there may be a sphere touching the tip of the **H** vector. Then, the wavelength whose reciprocal is the radius of this sphere is diffracted (Figure 4.13). A real crystal has a number of planes of different orientation and spacing. There are also various **H** vectors corresponding to these planes. When an X-ray beam of fixed wavelength is incident onto the crystal at a specific angle, diffraction occurs by the **H** vector that terminates on the surface of the Ewald sphere, as shown in Figure 4.14. It means that only a specific plane satisfies the Bragg condition in this case. The Bragg law is valid for any incident beam that makes an angle θ_B with the reflecting plane. The beam can be incident in any directions once this Bragg angle is maintained. Thus, the incident and diffracted beams constitute a cone centered about the **H** vector (Figure 4.15). When the incident beam lies on a cone of semi-angle $90 - \theta_B$ about the normal to the (*hkl*) plane, the diffracted beam also lies on the same cone in order to be coplanar with the normal to the plane and the incident beam. In the Ewald construction, the origin of the reciprocal lattice (i.e., the origin of the **H** vector) is always placed at the tip of the incident beam vector. This makes incident and diffracted cones separately generated in the Ewald sphere. Whenever the X ray beam is incident along the side of the incident cone, it can be diffracted along the side of the diffracted cone. Of course, diffraction occurs in such a direction that the incident beam, diffracted beam, and **H** vector are coplanar.

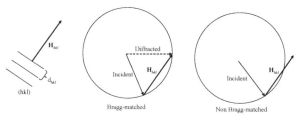

FIGURE 4.12 Diffraction condition when the reflecting plane and the X-ray wavelength are both fixed. The X-ray beam should be incident at a specific angle. For other angles of incidence, the corresponding $\mathbf{H}_{hkl}$ vector cannot terminate on the surface of the Ewald sphere.

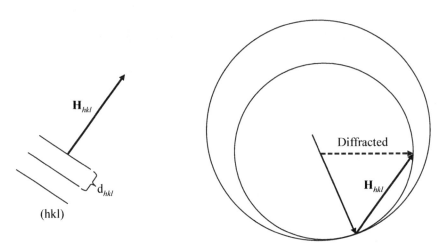

FIGURE 4.13 Diffraction condition when white X-rays are incident into the fixed reflecting plane at a specific angle.

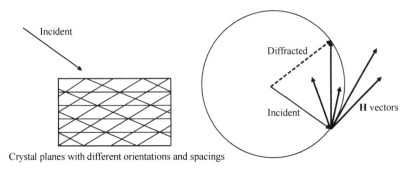

FIGURE 4.14 Diffraction condition when an X-ray beam of fixed wavelength is incident onto the crystal at a specific angle. Among many **H** vectors corresponding to different sets of crystal planes, diffraction occurs by the **H** vector that terminates on the surface of the Ewald sphere.

Directions of X-ray Diffraction

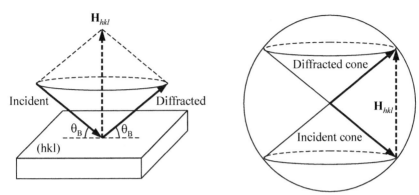

FIGURE 4.15 Incident and diffracted cones.

EXAMPLE 4.3

White X-rays of $\lambda = 1.5–1.6$ Å are incident on (001)-oriented sample **A** at a fixed angle of 35° and the beam diffracted from the sample **A** is made incident onto (111)-oriented sample **B**, as shown in Figure 4.16. The angle θ can be varied by rotating the sample **B** and the detector. If θ changes from 20° to 70°, how many diffraction peaks would be observed? The sample **A** has a simple cubic lattice ($a = 1.35$ Å) and **B** is simple tetragonal with $a = 3.0$ Å and $c = 3.6$ Å.

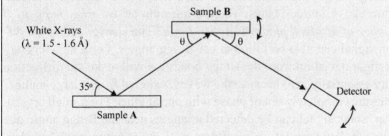

FIGURE 4.16 Diffraction geometry.

Answer:

Because the plane orientation and the incident angle were fixed, the sample **A** diffracts only a specific wavelength. We know that the wavelength of the diffracted beam is 1.548 Å from the equation of $\lambda = 2 \times 1.35$Å $\times \sin35°$. Since the sample **B** is (111)-oriented, diffraction can occur from (111),

> (222), (333), etc., planes. The spacing of these planes can be calculated using Eq. (2.23); $d_{111} = 1.828$ Å, $d_{222} = 0.914$ Å, and $d_{333} = 0.609$ Å. No diffraction peak will be observed from the (333) planes because d_{333} does not meet the fundamental requirement of $\sin\theta = \lambda/2d < 1$. Thus, we have 001 and 002 reflections only, whose angular positions are given below.
>
> $$1.548 \text{ Å} = 2 d_{111} \times \sin\theta, \; \theta = 25.02°$$
> $$1.548 \text{ Å} = 2 d_{222} \times \sin\theta, \; \theta = 57.78°$$

4.5 OFF-BRAGG ANGLE DIFFRACTION

In Section 3.3, we have considered a linear array of N scattering centers and investigated the dependence of diffraction intensity on the phase difference δ between scattered waves. The intensity was essentially zero except for the limited range of $2\pi - 2\pi/N < \delta < 2\pi + 2\pi/N$. This means that the condition for constructive interference becomes stricter with more scattering centers. Conversely, it states that the diffraction condition is mitigated as the number of involved scattering centers decreases. X-ray diffraction is a consequence of the constructive interference between waves scattered from successive lattice planes. Then, a question arises; *what happens if the number of involved planes is not so large?* The answer is that the diffraction signal can also be found at off-Bragg angles. For a large crystal, only a slight deviation from the Bragg condition will make the diffraction intensity essentially zero because the waves scattered from a large number of planes are completely out of phase with one another. For a small crystal, however, some signal can be detected at angles near the Bragg angle due to the incomplete destructive interference. As the resulting peak broadening can provide a method for estimating the size of small crystals, it is worthwhile to examine the scattering of X-rays incident at angles deviating from the Bragg angle.

Directions of X-ray Diffraction

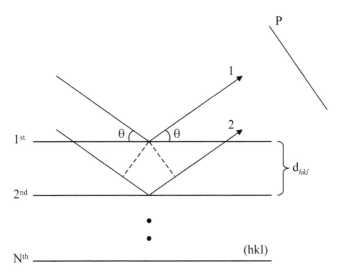

FIGURE 4.17 Diffraction from a thin crystal with a total of N planes.

Suppose that a crystal has a thickness t in the direction perpendicular to a particular set of (hkl) planes and contains a total of N planes of spacing d_{hkl} (Figure 4.17). When the X-rays of wavelength λ is incident onto the planes at an angle θ, ray **1** scattered from the first plane has a path length difference of $L = 2d_{hkl} \sin\theta$ with ray **2** scattered from the second plane. Any rays scattered from two adjacent planes has the same path length difference. The difference in phase between the two rays is then $\delta = 2\pi L / \lambda = 4\pi d_{hkl} \sin\theta / \lambda$. If the scattered rays are assumed to have the same amplitude E_o, the intensity I at some remote position P is then given by

$$I = E_o^2 \frac{\sin^2(N\delta/2)}{\sin^2(\delta/2)} \qquad (4.10)$$

Equation(4.10) was already derived in section 3.3.

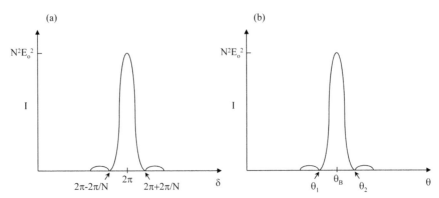

FIGURE 4.18 (a) Diffraction intensity vs. phase difference. (b) Diffraction intensity vs. scattering angle.

The dependence of intensity on the phase difference is plotted in Figure 4.18(a). When the incident angle is completely Bragg-matched (i.e., $\theta = \theta_B$), the path length difference L is equal to one wavelength and then, we have $\delta = 2\pi$ and $I = N^2 E_o^2$. As the phase difference δ deviates from 2π, the intensity I decreases and drops to zero at $\delta = 2\pi \pm 2\pi/N$. The diffraction intensity will also be lowered with an incident angle deviating from the Bragg angle, as shown in Figure 4.18(b). The angles θ_1 and θ_2 at which the intensity becomes zero can be derived from the following relations.

$$2\pi - 2\pi/N = 4\pi d_{hkl} \sin\theta_1 / \lambda$$

$$2\pi + 2\pi/N = 4\pi d_{hkl} \sin\theta_2 / \lambda \quad (4.11)$$

Side lobes outside the main peak are seldom observed in X-ray diffraction. From Eq. (4.11), we obtain

$$\sin\theta_2 - \sin\theta_1 = 2\cos(\frac{\theta_2 + \theta_1}{2})\sin(\frac{\theta_2 - \theta_1}{2}) = \frac{\lambda}{Nd_{hkl}} \approx 2\cos\theta_B \sin\{\frac{\theta_2 - \theta_1}{2}\} \quad (4.12)$$

The approximation $(\theta_2 + \theta_1)/2 \approx \theta_B$ is valid because θ_2 and θ_1 are almost equally away from θ_B. The angular separation, $\theta_2 - \theta_1$, can be calculated as follows.

$$\sin\left\{\frac{\theta_2 - \theta_1}{2}\right\} = \frac{\lambda}{Nd_{hkl} 2\cos\theta_B} \approx \frac{\theta_2 - \theta_1}{2}$$

Directions of X-ray Diffraction

$$\theta_2 - \theta_1 \approx \frac{\lambda}{Nd_{hkl}\cos\theta_B} \approx \frac{\lambda}{t\cos\theta_B} \quad (4.13)$$

where $(\theta_2 - \theta_1)/2$ is a small value compared to the absolute values of θ_2 and θ_1. Thus, $\sin\{(\theta_2 - \theta_1)/2\}$ can be approximated to $(\theta_2 - \theta_1)/2$. The approximation $t = (N-1)d_{hkl} \approx Nd_{hkl}$ is also acceptable unless the crystal has just a few reflecting planes. As stated earlier, 2θ value rather than θ is experimentally measured in X-ray diffraction. The broadening of a peak is customarily evaluated by measuring its intensity at half the maximum value (Figure 4.19).

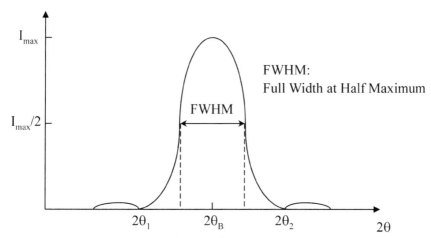

FIGURE 4.19 Definition of the Full Width at Half Maximum.

The measured peak width is referred to as the Full Width at Half Maximum (FWHM). As FWHM is approximately half the separation between two angles $2\theta_2$ and $2\theta_1$, it is expressed by Eq. (4.13). A more accurate analysis on this problem gives

$$FWHM \approx \theta_2 - \theta_1 = \frac{0.9\lambda}{t\cos\theta_B} \quad (4.14)$$

This is known as the **Scherrer equation** or formula. It states that the peak width is inversely proportional to the crystal thickness measured perpendicular to the reflecting planes. This means that a smaller crystal results in a more broadened peak. Consider a reflection from (001) planes of

$d_{001} = 3$ Å at $\lambda = 1.54$ Å. A crystal with $t = 1$ μm will have an FWHM value of $\sim 0.01°$, while it is increased to $\sim 0.2°$ for a 50 nm-thick crystal (Figure 4.20). When the phase difference δ is exactly 2π, complete constructive interference will result regardless of the magnitude of N. However, the situation for destructive interference is different. For a big crystal, waves scattered from a large number of reflecting planes annul one another even for a phase difference slightly different than 2π. On the contrary, a tiny crystal contains a much smaller number of planes. Thus, when the phase difference slightly deviates from 2π, the scattered waves do not completely annul one another. Assume $\delta = 2\pi + 2\pi/100$ as a specific example. According to Figure 4.18(a), a crystal with $N = 100$ has no diffraction at this phase difference but a smaller crystal with $N = 50$ will give a non-zero intensity.

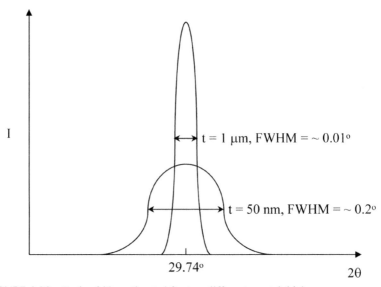

FIGURE 4.20 Peak widths estimated for two different crystal thicknesses.

The Scherrer equation is a formula that relates the size of sub-micrometer particles or crystallites in a solid to the broadening of a diffraction peak. It is often used to determine the size of crystalline particles in the form of powder. It is to be noted that the Scherrer formula is applicable to particles or grains smaller than about 0.2 μm and provides a lower bound on the particle size. The Scherrer equation is limited to nano-scale crystallites. It is not applicable to grains larger than about 0.2 μm, which precludes those observed in most metallographic microstructures. There are many other factors contributing to the broadening of a diffraction peak. In actual experiments, the incident beam is not perfectly parallel but contains convergent and divergent rays as well as parallel rays. In addition, the characteristic line (e.g., Cu K_α line) used as the monochromatic X-ray source is not perfectly monochromatic either, even after passing through a monochromator. Therefore, fairly large crystals of perfect crystalline quality will exhibit a non-zero peak width. Besides these instrumental effects, there are also material-related sources of the peak broadening, which include inhomogeneous strain, lattice imperfections, dislocations, and so on. If the other contributions to the peak width are non-zero, the actual crystallite size can be larger than that predicted by the Scherrer formula. As the particle size decreases, the size effect becomes more dominant than the other factors and the Scherrer formula would apply. For each *hkl* reflection, the value of *t* is interpreted as an average crystal dimension perpendicular to the reflecting planes.

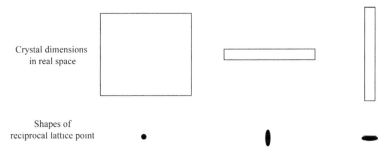

FIGURE 4.21 Relationship between the crystal dimensions in real space and the shapes of reciprocal lattice point.

The peak broadening caused by small crystal dimensions can also be related to the diffraction condition in reciprocal space. This reciprocal approach is very useful for explaining thin film X-ray diffraction and transmission electron microscopy. If a feature's dimension along a certain direction in real apace is small, the feature size along that direction in reciprocal space is large. The converse is also true. A crystal with infinitely large dimensions has very tiny reciprocal lattice points that have essentially no volume. When the crystal dimension is reduced along a certain direction, it reciprocal lattice points are elongated along that direction. Thus, a thin plate crystal will exhibit reciprocal lattice points of rod shape, while a large bulk crystal has small spherical lattice points (Figure 4.21). This rod shape is a consequence of the elongation of lattice points along the direction of the reduced dimension. As the crystal is more thinned, the rod becomes longer. Let's consider diffraction from a thin film whose surface is parallel to (001) planes, as shown in Figure 4.22. The sample is thin and the number of reflecting planes is small. It is thus expected that a diffraction peak exhibit quite a large width. This thin crystal has a reciprocal lattice composed of rods aligned perpendicular to the film surface. Then, the length of the H_{001} vector is not fixed but has a certain range. This enables diffraction to take place at other incidence angles near the Bragg angle θ_B, resulting in peak broadening. As the film gets thinner, the peak will be more broadened. When it is too thin, however, the overall diffraction intensity becomes very weak, making it difficult to define the peak width. It is important to note that the reciprocal lattice vector H may have a certain range in its direction as well as in the magnitude. Consider a thin crystal layer of cubic structure. If the layer has a small dimension in the z-direction, its reciprocal lattice points are elongated along the z-direction, as illustrated in Figure 4.23. Then, the H_{010} vector (also other vectors such as H_{020} and H_{100}) has divergent directions. In this case, the H_{001} vector has multiple lengths with a nearly fixed direction.

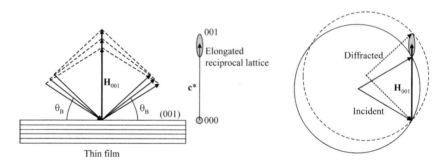

FIGURE 4.22 Peak broadening related to the diffraction condition in reciprocal space.

FIGURE 4.23 Origin of divergent H vector directions.

Directions of X-ray Diffraction

4.6 ELECTRON DIFFRACTION

The de Broglie postulate, formulated in 1924, predicts that particles should also behave like waves. De Broglie's hypothesis was confirmed some years later with the observation of electron diffraction in two independent experiments by G. Thomson, C. Davisson, and L. Germer. The wavelength of a traveling electron is given by the following de Broglie equation

$$\lambda = \frac{h}{m_o \upsilon} \tag{4.15}$$

where h is Plank's constant, and m_o, the rest mass of the electron. The electrons are accelerated in an electric potential V to the desired velocity $\upsilon = \sqrt{2eV / m_o}$. The electron wavelength is thus given by $\lambda = h / \sqrt{2m_o eV}$. Electron diffraction is usually carried out in a transmission electron microscope (TEM). Since the electrons in a TEM are accelerated to a velocity comparable to the speed of light, a relativistic modification should be made. The formula for the electron wavelength is then modified to

$$\lambda = h \left[2m_o eV \left(1 + \frac{eV}{2m_o c^2} \right) \right]^{-1/2} \tag{4.16}$$

where c is the speed of light. The electron wavelength in a 200 kV TEM is 0.025 Å (Cu K_α, $\lambda = 1.54$ Å). In a TEM analysis, the electrons pass through a thin film of the material to be investigated. The thin sample has reciprocal lattice points elongated along the direction of an incident electron beam (elongated reciprocal lattice points are hereafter referred to as *reciprocal lattice rods*). As mentioned above, the electron wavelength used in electron diffraction is much shorter than the wavelength of X-rays. This means that the radius of the Ewald sphere is much larger in electron diffraction experiments than in X-ray diffraction. As a consequence, the quite flat surface of the Ewald sphere can intersect a considerable number of reciprocal lattice rods, as shown in Figure 4.24.

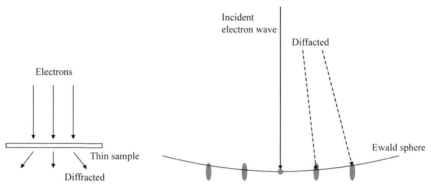

FIGURE 4.24 Principle of electron diffraction.

The radius, $1/\lambda$, of the Ewald sphere is of the order of 40 Å^{-1} in a typical TEM and the reciprocal lattice constants of crystals are of the order of 0.1 Å^{-1}. Diffraction occurs by reciprocal lattice vectors terminating on the surface of this sphere of reflection. Since the electrons have a very small wavelength, the diffraction angles are also much smaller than those of X-ray diffraction. Unlike X-ray diffraction, the electron diffraction arises from planes parallel to the incident beam. The resulting diffraction pattern is recorded on a fluorescent screen or photographic film (Figure 4.25). If the electron beam is incident along a zone axis of the crystal, the incident electrons are reflected from the planes that belong to this zone. That is, when the incident beam is parallel to the [uvw] direction of the crystal, it can be diffracted from the planes whose Miller indices hkl satisfy the relation of $hu + kv + lw = 0$ (refer to Eq. (2.19)). This allows a two-dimensional distribution of the reciprocal lattice to be revealed. While the diffracted beams have appreciable intensity at small diffraction angles, the intensity falls off rapidly as the diffraction angle increases. The reciprocal lattice rods are more unlikely to touch the Ewald sphere when they are further away from the reciprocal space origin. Thus, the spots in a diffraction pattern are from the planes with low Miller indices. The central spot is caused by a part of the beam that is not diffracted. Figure 4.26 shows TEM image of a Si film and its electron diffraction pattern. The indexing of diffraction spots will be discussed in the following chapter.

Directions of X-ray Diffraction 143

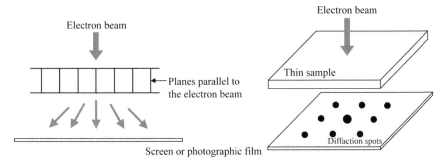

FIGURE 4.25 Formation of electron diffraction spots.

FIGURE 4.26 Transmission electron microscopy image of a Si film and electron diffraction pattern.

In TEM, a single crystal grain or particle may be selected for the diffraction experiments. If the sample is tilted with respect to the incident electron beam, one can obtain diffraction patterns from several crystal orientations. In this way, we can map the reciprocal lattice of the crystal in three dimensions. Although electron diffraction is a very powerful technique for analyzing the crystalline quality and orientation of a material, it is subject to several limitations. First of all, the sample to be studied should be electron transparent and be made very thin. The required thickness is usually less than 100 nm. Therefore, a careful and time-consuming sample preparation procedure is necessary. Furthermore, many samples are vulnerable to radiation damage induced by the incident electrons. The study of magnetic materials by electron diffraction is very complicated because electrons in magnetic fields are deflected by the Lorentz force. Although this effect may be utilized to study the magnetic domains of materials by Lorenz force microscopy, it makes structure determination virtually impossible. While both X-ray and neutron diffraction experiments are highly automated and routinely executed, electron diffraction requires a much higher level of user interaction. X-ray and neutron diffractions are therefore the preferred methods for determining lattice parameters and atomic positions.

PROBLEMS

4.1. A material of simple tetragonal Bravais lattice ($a = 3$ Å, $c = 2$ Å) is prepared in a plate shape as shown in Figure 4.27. An X-ray beam of 1.54 Å is incident at an angle of θ with the sample surface that is parallel to (001). What is the angle θ in order for diffraction to occur from $(0\bar{1}2)$ plane? If the wavelength of the incident X-ray beam increases by 0.1 Å, then how much should the incident angle change to get a diffraction peak from the same $(0\bar{1}2)$ plane?

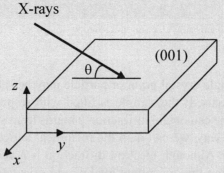

FIGURE 4.27.

4.2. Suppose that a white X-ray beam ($\lambda = 0.5$–3 Å) is incident on a (001)-oriented thin plate sample (simple cubic with $a = 2.0$ Å) at an angle of $\theta = 30°$ (Figure 4.28). The incident X-ray beam will be partially reflected as a result of diffraction and the rest will be transmitted. If we analyze the spectrum (wavelength vs. intensity) of the transmitted beam, the reduction in transmitted intensity will be observed at specific wavelengths.

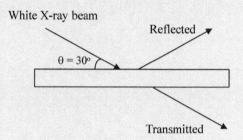

FIGURE 4.28 Reflection and transmission from a thin plate sample.

(a) At which wavelengths will the intensity reduction be observed? Assume that the incident beam has a uniform intensity distribution over the given wavelength range.

(b) How does the result change when the X-ray beam is normally incident, i.e., $\theta = 90°$?

4.3. State the advantages and limitations of electron diffraction, relative to X-ray diffraction.

4.4. The Bragg law $\lambda = 2d\sin\theta$ predicts only the possible directions of diffraction, but it mentions nothing about the diffraction intensity. In the cubic system, the interplanar spacing d is dependent on the lattice parameter only, not influenced by the crystal structure. Then, can the intensity of a diffraction signal predicted by the Bragg law be zero? If yes, explain why?

CHAPTER 5

INTENSITIES OF X-RAY DIFFRACTION

CONTENTS

5.1 Introduction 148

5.2 Scattering by an Electron 151

5.3 Scattering by an Atom 154

5.4 Scattering by a Unit Cell and Structure Factor 157

5.5 Systematic Absence 171

Problems 176

148   X-Ray Diffraction for Materials Research: From Fundamentals to Applications

5.1 INTRODUCTION

In the previous chapter, it has been shown that the diffraction direction is determined by the Bragg law: $\lambda = 2d\sin\theta$. It states that an X-ray beam of given wavelength can be diffracted from a given set of reflecting planes when it is incident at a specific angle called the Bragg angle. For a set of (*hkl*) planes, the diffraction angle $2\theta_{hkl}$ can be obtained from the relation of $\sin\theta_{hkl} = \lambda / 2d_{hkl}$. For any set of planes, the diffraction direction (i.e., the diffraction angle) is influenced only by the interplanar spacing. Figure 5.1 shows an example where the Bragg angle for (001) decreases as a result of the increase in d_{001}. Since the interplanar spacing d_{hkl} is dependent on the crystal system to which the crystal belongs and its lattice constants, the diffraction direction is solely determined by the shape and size of the unit cell. The crystal structures shown in Figure 5.1 are rather simple, with atoms only on the corners of the unit cell. Even though the crystal has a complex structure with many atoms within the unit cell, the diffraction angle for a specific set of (*hkl*) planes is not influenced by the presence of these atoms once the unit cell dimensions remain unaltered.

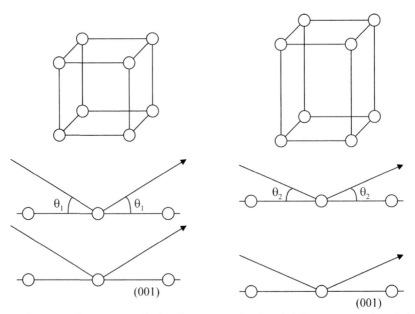

FIGURE 5.1   Decrease of the Bragg angle for (001) as a result of the increase in d_{001}.

It is important to note that the Bragg law mentions nothing about the intensity of a diffracted beam and suggests only the possible directions of diffraction. No diffraction signal may be observed in the direction predicted by the Bragg law. The zero intensity means that diffraction does not take place. The Bragg law is a necessary condition for diffraction to occur, but it is not a sufficient condition. Then, what determines the intensity of a diffracted beam? This is the subject that we will discuss in this chapter. Although there are many variables involved, the diffraction intensity is dominantly determined by the arrangement of atoms within the unit cell. As a simple example, consider 001 reflection from three different cubic crystals with identical lattice constant "a". Figure 5.2 shows simple cubic (top), body-centered cubic (middle), and CsCl (bottom) structures. Since all these structures are assumed to have the same lattice constant (i.e., d_{001} value), their Bragg angles for 001 reflection, θ_{001}, are also identical. Thus, an X-ray beam should be incident at this angle in all cases. It should be noted that this is the minimum requirement for diffraction and that the diffraction intensity depends on the actual crystal structure. When the crystal has a simple cubic structure, the path difference between rays **1** and **2** is one wavelength and diffraction occurs in the direction shown. In a crystal of body-centered cubic structure, rays **1** and **2** are also in phase with each other. In this case, however, there is another plane of atoms midway between the (001) planes. Since the path difference between rays **1** and **3** is one-half wavelength, they are completely out of phase and annul each other. Similarly, ray **4** from the next plane (not shown) annuls ray **2**. It means that the diffraction intensity from the set of (001) planes is zero, even though the Bragg condition is satisfied. In other words, there is no 001 reflection from the body-centered cubic structure. In the CsCl structure, rays **1** and **3** are also $180°$ out of phase. However, they are scattered from different types of atoms. As will be discussed later in this chapter, different atoms have different scattering strengths. Since the rays **1** and **3** have unequal amplitudes, complete destructive interference will not result. Thus, we have non-zero diffraction intensity in the given direction. Of course, the diffracted beam will be much weaker compared to the 001 reflection from a simple cubic structure of the same lattice constant.

150   X-Ray Diffraction for Materials Research: From Fundamentals to Applications

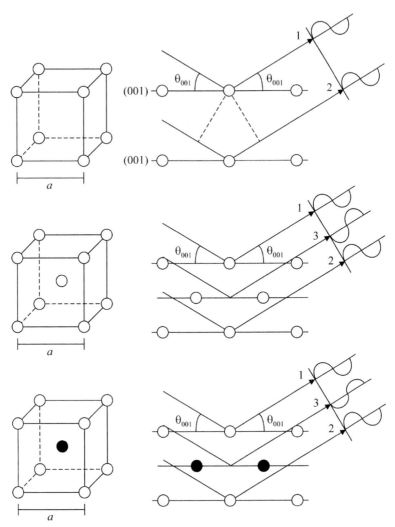

FIGURE 5.2   Diffraction from three different cubic structures (top: simple cubic structure, middle: body-centered cubic structure, bottom: CsCl structure). When these structures have the same lattice constant, their Bragg angles for 001 reflection, θ_{001}, are also identical. However, the diffraction intensity is significantly affected by the arrangement of atoms within the unit cell.

The above example shows that the intensity of a diffraction signal predicted by the Bragg law is significantly affected by the actual arrangement of atoms within the unit cell. Conversely, it also indicates that we can deduce atomic positions within the unit cell by measuring the diffraction intensities. The Bragg condition in reciprocal space is graphically illustrated in Figure 4.9. It states that in order for a monochromic X-ray beam to reflect from a set of (*hkl*) planes, the difference between the scattered and incident beam vectors should be equal to the reciprocal lattice vector $\mathbf{H}_{hkl}$. As we have already seen above, the incident beam may not be diffracted even though the Bragg condition is satisfied. It arises from the fact that the corresponding reciprocal lattice point is missing, which is known as the *systematic absence*. In this respect, the reciprocal lattice of a crystal should be constructed taking the systematic absence into account. In order to discuss all issues relevant to the diffraction intensity, we need to consider scattering by an electron at first.

5.2 SCATTERING BY AN ELECTRON

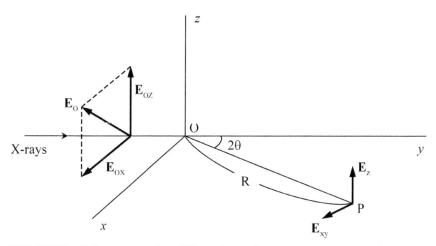

FIGURE 5.3 Coherent scattering of X-rays by an electron.

The X-ray beam is an electromagnetic wave with electric field varying sinusoidally with time at any position and directed perpendicular to the direction of propagation. This electric field exerts forces on the electrons of atom, making them oscillate about a mean position. According to the classical electromagnetic theory, an oscillating charge radiates, i.e., emits an electromagnetic wave. This radiation has the same frequency and wavelength as the primary beam and is called scattered radiation. Although X-rays are scattered in all directions, the intensity of the scattered radiation depends on the angle of scattering. Let us consider Figure 5.3, in which an X-ray beam is incident along the y-axis into an electron located at the origin O. We are interested in the scattered intensity at P, which is at a distance R from the electron at an angle 2θ with the y-axis. The x- and z-axes are chosen so that the point P is in the x-y plane. Since the primary beam generated from an X-ray tube is unpolarized, its electric field vector $\mathbf{E}_o$ is in a random direction in the x-z plane and is expressed as

$$\mathbf{E}_o = \hat{\mathbf{u}}E_o \cos(2\pi vt) \tag{5.1}$$

where E_o is the amplitude of a time-varying electric field and $\hat{\mathbf{u}}$ is a unit vector along the field. Since $\mathbf{E}_o$ is a vector, it can be resolved into two components $\mathbf{E}_{ox}$ and $\mathbf{E}_{oz}$. Thus, we obtain the following relations.

$$\mathbf{E}_o = \mathbf{E}_{ox} + \mathbf{E}_{oz} = \mathbf{i}E_{ox} \cos\left(2\pi vt\right) + \mathbf{k}E_{oz} \cos\left(2\pi vt\right) \tag{5.2}$$

where E_{ox} and E_{oz} are the amplitudes of the component fields. The amplitude E of the scattered electric field $\mathbf{E}$ at distance R from the origin is given by

$$E = E_o \frac{\mu_o e^2}{4\pi mR} \sin\alpha \tag{5.3}$$

where μ_o is the magnetic susceptibility in vacuum ($4\pi \times 10^{-7}$ m kg C^{-2}) and α is the angle between the scattering direction and the oscillation direction of the electron. The $\mathbf{E}_{oz}$ component of the primary beam oscillates the electron along the z-axis, giving rise to a scattered field $\mathbf{E}_z$ at P. Its amplitude is found from Eq. (5.3) to be

$$E_z = E_{oz} \frac{\mu_o e^2}{4\pi mR} \tag{5.4}$$

Similarly, the amplitude of the scattered component $\mathbf{E}_{xy}$ is given by

$$E_{xy} = E_{ox} \frac{\mu_o e^2}{4\pi mR} \cos(2\theta) \tag{5.5}$$

Intensities of X-ray Diffraction 153

Here, $\sin \alpha = \sin\left(\frac{\pi}{2} - 2\theta\right) = \cos(2\theta)$. The resultant amplitude at P is then given by

$$E^2 = E_z^2 + E_{xy}^2 = \left(\frac{\mu_o e^2}{4\pi m R}\right)^2 (E_{oz}^2 + E_{ox}^2 \cos^2 2\theta) \qquad (5.6)$$

The amplitudes of the two components $\mathbf{E}_{ox}$ and $\mathbf{E}_{oz}$ will be identical on the average because the electric field vector $\mathbf{E}_o$ of the primary beam takes all orientations in the x–z plane with equal probability; $E_{oz}^2 = E_{ox}^2 = E_o^2 / 2$. This leads to

$$E^2 = E_o^2 \left(\frac{\mu_o e^2}{4\pi m R}\right)^2 \left(\frac{1 + \cos^2 2\theta}{2}\right)$$

$$I = I_o \left(\frac{\mu_o e^2}{4\pi m R}\right)^2 \left(\frac{1 + \cos^2 2\theta}{2}\right) \qquad (5.7)$$

where I and I_o are the intensities of the scattered and primary beams, respectively. Eq. (5.7) represents the intensity of classical scattering by a single electron and is known as the *Thomson equation*. The factor $(1 + \cos^2 2\theta) / 2$ is called the polarization factor for an unpolarized primary beam. If the primary beam is polarized, this factor takes a different form. For R of about 1 cm, I/I_o is of the order of 10^{-26}. One might think that the intensity of scattered X-rays is too small to measure. But we need to recall that only 1 mg of matter contains approximately 10^{20} electrons. The intensity of X-rays scattered from a sample is not too low to measure. It can be considerable when the scattered waves constructively interfere in particular directions. As shown in Eq. (5.7), the scattering intensity is stronger in forward or backward directions than at right angles to the incident beam. Since the scattering intensity increases in proportion to the intensity of the incident beam, its absolute value is not so meaningful. Relative values are sufficient for most X-ray diffraction experiments.

In the above-mentioned *coherent scattering*, there is a definite phase relationship between the scattered beam and the incident beam. Since the phase change on coherent scattering is identical for all the electrons in a material, we don't need to consider it in deriving the diffraction condition. The coherently scattered X-rays have the same wavelength and frequency as the incident X-ray beam. The interference between such coherently scattered rays is the basis of X-ray diffraction. The scattering of X-rays by an electron may occur in a quite different way. An X-ray photon encountered with a loosely bound or free electron can be deflected by the electromagnetic field of the electron and give some of its energy to the electron as kinetic energy. It is like the collision between two billiard balls. Thus, the deflected (i.e., scattered) X-ray photon has lower energy than the incident photon. As a result, the wavelength of the scattered radiation is slightly longer than that of the incident beam. This effect, discovered by A. Compton, is called the *Compton scattering* or *effect*. The Compton scattering can be understood only by considering the incident beam as a stream of X-ray photons (quanta), and it cannot be explained with the wave theory. It is important to note that the phase of the Compton-scattered radiation has no fixed relation to the phase of the incident beam. This *incoherent* radiation cannot participate in X-ray diffraction because interference does not take place between the waves at random phases. The incoherently scattered rays have the undesirable effect of increasing the background level in diffraction patterns.

5.3   SCATTERING BY AN ATOM

An atom of atomic number Z contains a positively charged nucleus and Z electrons. When an X-ray beam encounters an atom, all of its electrons scatter part of the incident radiation in accordance with the Thomson equation. However, the nucleus makes a negligible contribution to the scattering process because it has a significantly large mass compared to that of electrons and is unable to respond with a rapidly-oscillating electric field. Therefore, the net scattering effect is only due to the electrons and the wave scattered by an atom is simply the sum of the waves scattered by its component electrons. Since the electrons of an atom are located at different positions, the amplitude of the wave scattered by an atom is determined by the phase relationship between the waves scattered by different electrons. Consider Figure 5.4, in which the electrons are situated at different positions around the nucleus. The amplitude of the wave scattered by an atom containing Z electrons becomes Z times the amplitude of the wave scattered by a single electron in the scattering of forward direction ($2\theta = 0$), because the waves scattered by all the electrons are in phase. This is not true for other scattering directions. For instance, the waves scattered in the forward direction by electrons on points A and B have no phase difference on a plane marked as X-X'. However, the waves scattered in a direc-

Intensities of X-ray Diffraction 155

tion shown in the figure have a path difference of $CB - AD$ and are thus somewhat out of phase on $Y\text{-}Y'$. This leads to partial interference between the scattered waves so that the resulting amplitude is less than that of the wave scattered in the forward direction. The *atomic scattering factor, f,* is defined as the ratio of the amplitude of the wave scattered in a particular direction by an atom to the amplitude of the wave scattered in the same direction by an electron. When the amplitude E of the wave scattered by a single electron is given by Eq. (5.7), *the amplitude E_a of the wave scattered by an atom is thus $E_a = fE$.* The atomic scattering factor *f* means the amplitude of scattering per atom expressed in units of the amplitude from a single electron.

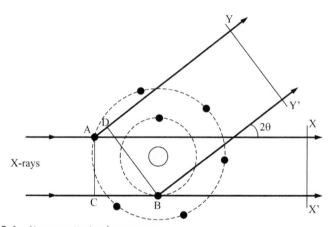

FIGURE 5.4   X-ray scattering by an atom.

In Figure 5.4, the electrons are depicted as situated at specific points. In reality, the electrons are spread out in space and occupy a finite volume. For an object that is spherically symmetric, the spatial density distribution can be expressed as a function of radius. If the electron distribution in an atom has spherical symmetry, each small element of volume dV within the electron cloud will give rise to scattered radiation of amplitude proportional to $\rho(r)dV$, where $\rho(r)$ is the electron density at a distance r from the center of the atom. For each atom, we have $\int \rho(r)dV = Z$. There will be a difference in path length between the waves scattered by any pair of volume elements; the path difference is zero for $2\theta = 0$ and will increase with increasing 2θ. To obtain the total amplitude of scattering from an atom, we should consider the phase relationship between all the contributing

156   X-Ray Diffraction for Materials Research: From Fundamentals to Applications

elements and integrate over the volume occupied by the electrons. When the distribution of electrons is spherically symmetric around a nucleus, the atomic scattering factor is given by

$$f = 4\pi \int_0^\infty \rho(r) r^2 \frac{\sin kr}{kr} dr \qquad (5.8)$$

where $k = 4\pi \sin\theta / \lambda$. This atomic scattering factor plays a significant role in X-ray diffraction. Atomic scattering factors are used to calculate the structure factor for a given Bragg peak. For any atom, it is a function of $\sin\theta/\lambda$, where θ is half the scattering angle and λ, the X-ray wavelength. When the scattering occurs in the forward direction, i.e., the scattering angle $2\theta = 0$, the above Eq. (5.8) reduces to

$$f = 4\pi \int_0^\infty \rho(r) r^2 dr = Z \qquad (5.9)$$

It is evident from Eq. (5.8) that f approaches Z at small values of $\sin\theta/\lambda$. As θ increases, the waves scattered by individual volume elements become more and more out of phase and f decreases. The atomic scattering factor also depends on the wavelength of the incident X-ray beam. At a fixed value of θ, the path difference will be larger as the wavelength gets shorter. Therefore, the atomic scattering factor will decrease with decreasing wavelength. To obtain f, we need only to know the radial distribution of the electron density in the atom. The atomic scattering factors calculated for various atoms are tabulated in Ref. [2, 3]. Those of some atoms are plotted in Figure 5.5. The atomic scattering factor is sometimes called the atomic form factor because it depends on how the electrons are distributed around the nucleus. The above treatment is rather simplified, and the atomic scattering factor is generally complex. However, the use of its real component is sufficient for ordinary X-ray diffraction since the imaginary components only become large near an absorption edge. Anomalous X-ray scattering makes use of the variation of the atomic scattering factor close to an absorption edge to evaluate the scattering power of specific atoms in the sample. This anomalous scattering is out of the scope of this book.

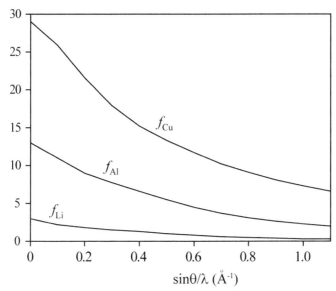

FIGURE 5.5 Atomic scattering factors of some elements.

5.4 SCATTERING BY A UNIT CELL AND STRUCTURE FACTOR

If the Bragg law is not satisfied, diffraction does not take place. Even if the Bragg law is satisfied for a particular set of atomic planes, no diffraction may occur, as discussed earlier in this chapter. The atoms of a crystal are periodically arranged with a repeat unit called the unit cell. The fact that no diffraction signal is observed even under the Bragg condition means that the waves scattered by all the atoms within the unit cell are completely out of phase with one another. The arrangement of atoms is identical in all unit cells. Thus, if complete destructive interference occurs between the waves scattered by a single unit cell, we have no diffracted beam because the crystal is just a repetition of the unit cell. In other words, the phase relationship between the waves scattered by the individual atoms of a unit cell is a dominant factor affecting the diffraction intensity.

158   X-Ray Diffraction for Materials Research: From Fundamentals to Applications

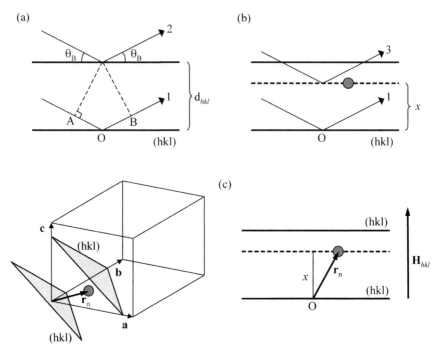

FIGURE 5.6 Effect of atom position on the phase difference between scattered rays. When the Bragg law is satisfied for a set of (hkl) planes, the phase difference between rays 1 and 2 is 2π. Therefore, rays 1 and 3 have a phase difference of $\delta = (x/d_{hkl})2\pi$.

Suppose that the Bragg law is satisfied for a set of (hkl) atomic planes when an X-ray beam of wavelength λ is incident at an angle θ_B, as shown in Figure 5.6(a). Here we take point O as the origin of the unit cell. The path length difference between scattered rays 1 and 2 is $AO + OB = \lambda = 2d_{hkl}\sin\theta_B$. Assume that there is another plane of atoms at a vertical distance x from the origin and ray 3 is reflected from this plane given by broken lines in Figure 5.6(b). Then, the path difference between the rays 1 and 3 will be $(x/d_{hkl})\lambda$. Since the phase difference between the rays 1 and 2 is 2π, the rays 1 and 3 have a phase difference of $\delta = (x/d_{hkl})2\pi$. Consider an atom on the broken-line plane whose position within the unit cell is represented by vector $\mathbf{r}_n$ in Figure 5.6(c). The position vector can be written as $\mathbf{r}_n = u\mathbf{a} + v\mathbf{b} + w\mathbf{c}$, where u, v, and w are the coordinates of the atom. We wish to know the phase difference between rays scattered by this atom and an atom situated at the origin O (not shown in the figure).

Intensities of X-ray Diffraction 159

As illustrated in Figure 4.2, scattered rays from any points of a plane are in phase with one another; there is no phase difference between the ray 3 and the ray scattered by the atom at r_n. Thus, the phase difference between the rays scattered by these two atoms will also be $\delta = (x/d_{hkl})2\pi$. The value of x can be obtained by forming the scalar product between r_n and $\mathbf{H}_{hkl}/\mathrm{H}_{hkl}$. The phase difference is then given by

$$\delta = \frac{2\pi x}{d_{hkl}} = \frac{2\pi}{d_{hkl}} \frac{\mathbf{H}_{hkl}}{\mathrm{H}_{hkl}} \cdot \mathbf{r}_n = 2\pi(hu + kv + lw) \qquad (5.10)$$

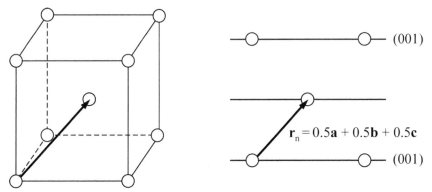

FIGURE 5.7 A body-centered atom represented by vector $r_n = 0.5\mathbf{a} + 0.5\mathbf{b} + 0.5\mathbf{c}$.

When the Bragg condition is satisfied for (hkl) planes, two atoms with coordinates (0,0,0) and (u,v,w) has a phase difference given by Eq. (5.10). As shown in Figure 5.7, a BCC crystal contains two equivalent atoms within the unit cell, each at (0,0,0) and (1/2,1/2,1/2). When the Bragg condition is satisfied for (001) planes, these two atoms have a phase difference of $\delta = \pi$. If the wave scattered by the corner atom is represented by $f\cos(2\pi x/\lambda)$, the scattered wave from the body-centered atom can be given by $f\cos(2\pi x/\lambda - \pi)$, where f is the atomic scattering factor of the atom. Since the total wave scattered from a unit cell is $f\cos(2\pi x/\lambda) + f\cos(2\pi x/\lambda - \pi) = 0$, diffraction does not occur from the (001) planes. Namely, there is no 001 reflection peak. One might think that we will have a diffracted beam when the path length difference between successive (001) planes is made equal to 2λ by increasing the incident angle. Note that this is the Bragg condition for (002) planes, not (001) planes; $\lambda = d_{001} \sin\theta = 2d_{002} \sin\theta$. Evidently, we have a 002 reflection peak from a BCC crystal because the phase difference between the same atoms is now $\delta = 2\pi$ and the scattered wave represented by $f\cos(2\pi x/\lambda) + f\cos(2\pi x/\lambda - 2\pi) = 2f\cos(2\pi x/\lambda)$ has non-zero amplitude.

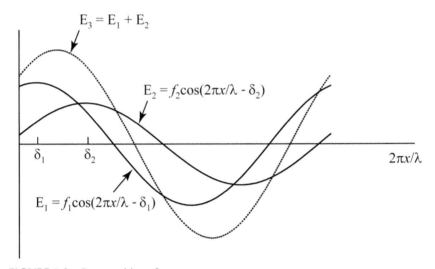

FIGURE 5.8 Superposition of waves.

It is mathematically simpler to manipulate the superposition of waves by use of complex number notation. A complex number with real and imaginary parts, for example, $a + ib$, is marked as a point in the complex plane and is represented by a vector drawn from the origin to this point. The length of this vector is the magnitude of the complex number. Figure 5.8 shows two waves $E_1 = f_1 \cos(2\pi x/\lambda - \delta_1)$ and $E_2 = f_2 \cos(2\pi x/\lambda - \delta_2)$. These waves have the same wavelength but differ in amplitude and phase. Their sum $E_3 = E_1 + E_2$ is given by the dotted curve. To obtain the intensity of the resultant wave E_3, we should know its amplitude. In this case, the amplitude and phase of E_3 can be easily calculated by trigonometric manipulation. However, the trigonometric calculation is obviously cumbersome when a larger number of waves are involved. From the relation of $Ae^{i\theta} = A\cos\theta + iA\sin\theta$, we find that a sinusoidal wave can be expressed in a complex exponential form. The complex number $Ae^{i\theta}$ is represented in the complex plane by a vector of length A inclined at an angle θ to the

Intensities of X-ray Diffraction   161

real axis. As θ increases, this vector is counterclockwise rotated and the real part of the complex number changes between A and $-A$. Likewise, the two waves shown in Figure 5.8 can be expressed in complex exponential forms.

$$E_1 = f_1 e^{i(2\pi x/\lambda - \delta_1)} = f_1 e^{-i\delta_1} e^{i(2\pi x/\lambda)}$$
$$E_2 = f_2 e^{i(2\pi x/\lambda - \delta_2)} = f_2 e^{-i\delta_2} e^{i(2\pi x/\lambda)} \quad (5.11)$$

The term $e^{i(2\pi x/\lambda)}$ is a common factor in all expressions and need not be considered anymore; it is here set to a unity for simplicity. The complex numbers E_1 and E_2 are represented by vectors $\mathbf{E}_1$ and $\mathbf{E}_2$ in Figure 5.9. The length of the vector is equal to the amplitude of the corresponding component wave. The amplitude f_3 and phase δ_3 of the resultant wave E_3 can be found simply by adding the vectors $\mathbf{E}_1$ and $\mathbf{E}_2$. As a practical example, Figure 5.10 shows how the amplitude and phase of the total wave can be determined, when the waves scattered by two different atoms has a certain phase difference.

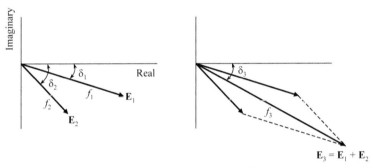

FIGURE 5.9 Complex representation of the superposition of waves with different amplitudes and phases.

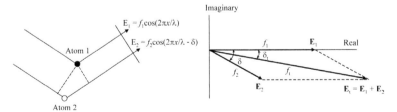

FIGURE 5.10 Superposition of the waves scattered by two different atoms.

We now return to the matter of adding the waves scattered by each of the atoms within the unit cell. The amplitude of each wave is given by the atomic scattering factor of the atom scattering it. As shown by Eq. (5.10), the phase of each wave is related to the Miller indices of the reflecting planes and the coordinates of the scattering atom. If a unit cell contains a total of N atoms, the resultant wave scattered by all the atoms of the unit cell will be given by

$$F = f_1 e^{i\delta_1} + f_2 e^{i\delta_2} + \ldots + f_N e^{i\delta_N} \tag{5.12}$$

where F is called the *structure factor*. The magnitude $|F|$ of the structure factor shown in Figure 5.11 represents the amplitude of the resultant wave. For hkl reflection, the structure factor is more concisely expressed as

$$F_{hkl} = \sum_{n=1}^{N} f_n e^{2\pi i (hu_n + kv_n + lw_n)} \tag{5.13}$$

where f_n is the atomic scattering factor of the nth atom and u_n, v_n, and w_n represent its coordinates. $|F|$ is defined as the ratio of the amplitude of the wave scattered by all the atoms of a unit cell to the amplitude of the wave scattered by a single electron. When the Bragg law is satisfied for a set of (hkl) planes, the diffraction intensity from these planes is proportional to $|F_{hkl}|^2$. The structure factor may be real or a complex number. For a complex structure factor F, the squared magnitude can be obtained with multiplication by its complex conjugate F^*; $|F|^2 = FF^*$. The structure factors for hkl and $\overline{hkl}$ are complex conjugates to each other; i.e., $F_{hkl} = F_{\overline{hkl}}^*$. This result means that the corresponding diffraction intensities are equal: $I_{hkl} = I_{\overline{hkl}}$. Thus, the diffraction pattern of a crystal is centrosymmetric irrespective of whether the crystal itself has a center of symmetry or not. If the structure factor F_{hkl} is zero for a certain reflection hkl, the intensity of that reflection will be zero. In this respect, Eq. (5.13) is a very important relation in X-ray diffraction (also in electron diffraction). Actual examples are not only helpful but also essential for understanding the meaning of a new equation or relation. The structures of many different crystals have been investigated in Section 2.3. The structure factors of some crystals are here calculated for this purpose.

Intensities of X-ray Diffraction 163

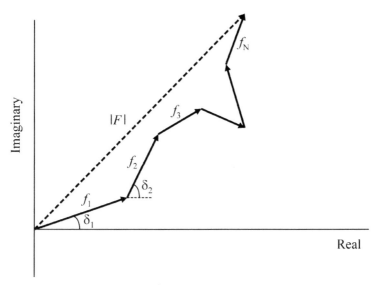

FIGURE 5.11 The magnitude $|F|$ of the structure factor represents the amplitude of the resultant wave.

(a) A crystal of FCC structure has four atoms of the same kind at (0,0,0), (1/2,1/2,0), (1/2,0,1/2), and (0,1/2,1/2) in a unit cell. The structure factor for *hkl* reflection is then

$$F = f[1 + e^{\pi i(h+k)} + e^{\pi i(h+l)} + e^{\pi i(k+l)}] \quad (5.14)$$

If *h*, *k*, and *l* are unmixed, i.e., all even or all odd, the three sums $(h + k)$, $(h + l)$, and $(k + l)$ are even integers and each exponential term in the above equation becomes 1. If *h*, *k*, and *l* are mixed, $e^{\pi i(h+k)} + e^{\pi i(h+l)} + e^{\pi i(k+l)} = -1$ regardless of whether two indices are even and one odd, or two odd and one even. The structure factor of the FCC structure can be summarized as

$$\begin{aligned} F &= 4f \quad \text{for unmixed } hkl \\ F &= 0 \quad \text{for mixed } hkl \end{aligned} \quad (5.15)$$

Thus, diffraction occurs for such planes as (111), (200), and (220) but not for (100), (110), (210), etc. When *h*, *k*, and *l* are unmixed, all the four atoms in the unit cell are in phase with one another. Therefore, their atomic scattering factors add up, resulting in $F = 4f$ as shown in Figure 5.12(a).

On the contrary, when h, k, and l are mixed, two atoms that are in phase with each other are 180° out of phase with the other two atoms. This leads to $F = 0$ (Figure 5.12(b)).

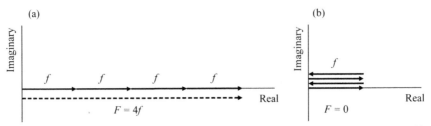

FIGURE 5.12 Complex-plane representation of the structure factor in FCC structure. (a) When the Miller indices h, k, and l are unmixed, all the four atoms in the unit cell are in phase and their atomic scattering factors add up, resulting in $F = 4f$. (b) For mixed h, k, and l, two atoms in phase are 180° out of phase with the other two atoms. This leads to $F = 0$.

(b) The CsCl structure contains two atoms per unit cell: one Cs at (0,0,0) and one Cl at (1/2,1/2,1/2). The structure factor is then given by

$$F = [f_{Cs} + f_{Cl}e^{\pi i(h+k+l)}]$$
$$F = (f_{Cs} + f_{Cl}) \text{ for } h+k+l = \text{even}$$
$$F = (f_{Cs} - f_{Cl}) \text{ for } h+k+l = \text{odd} \quad (5.16)$$

When $(h + k + l)$ is even, scattered rays from the Cs and Cl atoms are in phase and $F = (f_{Cs} + f_{Cl})$ is obtained, as shown in Figure 5.13(a). Although the scattered rays are completely out of phase when $(h + k + l)$ is odd, the structure factor does not become zero because Cs and Cl have different atomic scattering factors (Figure 5.13(b)). The BCC structure has two atoms of the same kind per unit cell located at (0,0,0) and (1/2,1/2,1/2). Thus, its structure factor can be easily calculated from Eq. (5.16) and takes the forms

$$F = 2f \quad \text{for } h+k+l = \text{even}$$
$$F = 0 \quad \text{for } h+k+l = \text{odd} \quad (5.17)$$

Intensities of X-ray Diffraction 165

In Section 5.1, we have shown with geometric analysis that there would be no 001 reflection from the BCC structure. This is consistent with the results of the structure factor calculation.

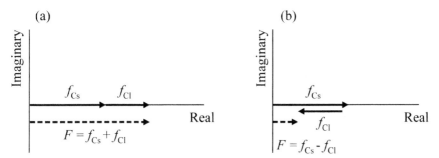

FIGURE 5.13 Complex-plane representation of the structure factor in CsCl structure. (a) $(h + k + l)$ is even, (b) $(h + k + l)$ is odd.

(c) We next consider the structure factor of NaCl structure. The unit cell of NaCl contains 4 Na and 4 Cl at the following positions:

Na: (0,0,0) (1/2,1/2,0) (1/2,0,1/2) (0,1/2,1/2)
Cl: (1/2,1/2,1/2) (0,0,1/2) (0,1/2,0) (1/2,0,0)

Inserting the atomic positions into Eq. (5.12) gives the structure factor as follows.

$$F = \left[1 + e^{\pi i(h+k)} + e^{\pi i(h+l)} + e^{\pi i(k+l)}\right]\left[f_{Na} + f_{Cl} e^{\pi i(h+k+l)}\right] \quad (5.18)$$

In the derivation of this equation, we made use of the relation of $e^{-n\pi i} = e^{n\pi i}$ where n is any integer. NaCl has a face-centered cubic lattice. The terms in the first bracket correspond to the face-centering translations and have already appeared in Eq. (5.14). They are 0 for mixed indices and 4 for unmixed indices. The terms in the second bracket describe the basis of a lattice point at the unit cell origin, namely, Na at (0,0,0) and Cl at (1/2,1/2,1/2). The structure factor of NaCl takes the three forms:

$$F = 4(f_{Na} + f_{Cl}) \quad \text{for all even } hkl$$
$$F = 4(f_{Na} - f_{Cl}) \quad \text{for all odd } hkl$$
$$F = 0 \quad \text{for mixed } hkl \quad (5.19)$$

166   X-Ray Diffraction for Materials Research: From Fundamentals to Applications

Like the FCC structure, the NaCl structure will exhibit diffraction peaks from the planes whose Miller indices are unmixed. However, the diffraction intensity obtained for all odd indices is much lower than that for all even indices.

(d) The zinc blende structure (sometimes called the sphalerite structure) also has a face-centered cubic lattice with two different atoms associated with one lattice point. In the zinc blende form of ZnS, the unit cell contains 4 Zn and 4 S at the following positions:

$$Zn: (0,0,0) \ (1/2,1/2,0) \ (1/2,0,1/2) \ (0,1/2,1/2)$$

$$S: (1/4,1/4,1/4) \ (3/4,3/4,1/4) \ (3/4,1/4,3/4) \ (1/4,3/4,3/4)$$

The structure factor will also contain the terms corresponding to the face-centering translations. We already know that they are zero for mixed indices. The lattice point at the origin is associated with Zn at $(0,0,0)$ and S at $(1/4,1/4,1/4)$. Thus, the structure factor for unmixed indices is given by

$$F = 4\left[f_{Zn} + f_S e^{\pi i (h+k+l)/2} \right] \tag{5.20}$$

Since it may be real or complex, we need to consider the squared magnitude:

$$|F|^2 = 16\left[f_{Zn} + f_S e^{\pi i (h+k+l)/2} \right]\left[f_{Zn} + f_S e^{-\pi i (h+k+l)/2} \right] = 16\left[f_{Zn}^2 + f_S^2 + 2 f_{Zn} f_S \cos\frac{\pi}{2}(h+k+l) \right]$$

$$\tag{5.21}$$

The structure factor of ZnS takes the four forms:

$$|F|^2 = 16\left(f_{Zn} + f_S \right)^2 \ \text{when} \ (h+k+l) \text{is an even multiple of 2}$$

$$|F|^2 = 16\left(f_{Zn} - f_S \right)^2 \ \text{when} \ (h+k+l) \text{is an odd multiple of 2}$$

$$|F|^2 = 16\left(f_{Zn}^2 + f_S^2 \right) \ \text{for all odd } hkl$$

$$|F|^2 = 0 \ \text{for mixed } hkl \tag{5.22}$$

In the diamond structure, the unit cell contains eight atoms of the same kind at the positions that would be occupied by Zn and S. Thus, $f_{Zn} = f_S$ and we have its structure factor given by

Intensities of X-ray Diffraction

$$|F|^2 = 64f^2 \text{ when } (h+k+l) \text{ is an even multiple of 2}$$
$$|F|^2 = 32f^2 \text{ for all odd } hkl$$
$$|F|^2 = 0 \text{ when } (h+k+l) \text{ is an odd multiple of 2}$$
$$|F|^2 = 0 \text{ for mixed } hkl \qquad (5.23)$$

Diamond has a face-centered cubic lattice. Therefore, diffraction does not occur from planes with mixed indices. Even for unmixed indices, however, reflections are missing when $(h+k+l)$ is an odd multiple of 2. This is because unlike other structures of FCC lattice, the diamond structure has two equivalent atoms associated with one lattice point (refer to Example 5.2). It is important to note the distinction between a structure and a Bravais lattice. NaCl, zinc blende, and diamond structures all have a face-centered cubic Bravais lattice. Of course, the Bravais lattice of FCC structure is also face-centered cubic. Although the structure factor mainly depends on the lattice type, it is also strongly influenced by the actual atomic arrangement as demonstrated above.

(e) The hexagonal close-packed structure shown in Figure 2.40(a) has two atoms of the same kind at $(0,0,0)$ and $(2/3,1/3,1/2)$. The structure factor is expressed as

$$|F|^2 = f^2 \left[1 + e^{2\pi i[\frac{2h+k}{3}+l/2]}\right]\left[1 + e^{-2\pi i[\frac{2h+k}{3}+l/2]}\right] = 4f^2 \cos^2 \pi(\frac{2h+k}{3}+\frac{l}{2}) \qquad (5.24)$$

It can be divided into the following four forms:

$$|F|^2 = 4f^2 \quad 2h+k = 3n, \ l = \text{even}$$
$$|F|^2 = 3f^2 \quad 2h+k = 3n\pm 1, \ l = \text{odd}$$
$$|F|^2 = f^2 \quad 2h+k = 3n\pm 1, \ l = \text{even}$$
$$|F|^2 = 0 \quad 2h+k = 3n, \ l = \text{odd} \qquad (5.25)$$

Here n is an integer. If $2h+k = 3n$, then $h+2k = 6n-3h = 3n'$. Thus, h and k are interchangeable in the condition of Eq. (5.25). The structure factor of the HCP structure can be recognized such that reflections are missing when $(2h+k)$ is a multiple of 3 and l is odd.

EXAMPLE 5.1

As shown in Figure 2.51, Cu_3Au is cubic with one Cu_3Au unit per unit cell. In the ordered form, the atomic positions are Au: (0,0,0) and Cu: (1/2,1/2,0), (1/2,0,1/2), and (0,1/2,1/2). In the disordered form, the same positions are randomly occupied and we can consider this random occupation equivalent to 1/4 Au and 3/4 Cu at each position. Derive the structure factors of Cu_3Au for the order and disorder forms and state for which *hkl* reflections the structure factor will be identical in the two forms.

Answer: From Eq. (5.13), the structure factor of the ordered form is given by

$$F = f_{Au} + f_{Cu}[e^{\pi i(h+k)} + e^{\pi i(h+l)} + e^{\pi i(k+l)}]$$

The structure factor is non-zero for any indices and all reflections are thus possible. This is due to the simple cubic Bravais lattice of the ordered Cu_3Au structure. In the disordered state, there is no preferred position for Cu or Au. The probability that a particular atomic site is occupied by Au atom is 1/4, the atomic fraction of Au in the alloy, and the probability that it is occupied by Cu is 3/4. Since every site has the same probability, the disordered form of Cu_3Au has an FCC structure on the average. Its structure factor is then equivalent to Eq. (5.14) and is expressed as

$$F = f + f[e^{\pi i(h+k)} + e^{\pi i(h+l)} + e^{\pi i(k+l)}]$$

where $f = f_{Au}/4 + 3f_{Cu}/4$. If h, k, and l are unmixed, each exponential term in the above equations becomes 1. Thus, both forms have the same structure factor of $F = f_{Au} + 3f_{Cu} = 4f$. For mixed indices, the structure factor of the disordered form is zero, while the ordered form has a non-zero value.

Intensities of X-ray Diffraction 169

EXAMPLE 5.2

Si and GaAs have the same Bravais lattice of FCC. Explain with geometrical considerations why 222 reflection is missing in Si, while it is observed from GaAs.

Answer: Si and GaAs have the diamond and zinc blende structures, respectively. It is evident from Eq. (5.22) and (5.23) that $F_{222} = 0$ in Si and $F_{222} \neq 0$ in GaAs. Figure 5.14(a) shows the stacking sequence of successive (111) planes along [111] in Si, which can be described as *A AB BC CA AB BC CA*. Successive planes are not equally separated from one another and there is another plane $d_{111}/4$ away from two planes with spacing of d_{111}. When the Bragg condition for 222 reflection is satisfied, the path length difference between rays **1** and **2** is 2λ. Since scattered rays **1** and **3** have a path difference of $\lambda/2$, they are completely out of phase and canceled out. Similarly, rays **2** and **4** annul each other. Thus, we have no diffraction signal. In GaAs, rays **1** and **3** are also 180° out of phase. However, they are scattered from different types of atoms and do not completely annul each other. Thus, an incident X-ray beam can be reflected in the given direction although the diffraction intensity may not be so high.

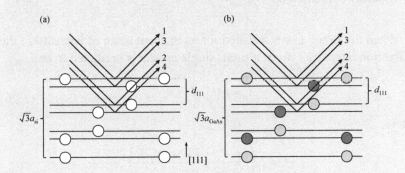

FIGURE 5.14 Diffraction in (a) Si and (b) GaAs when the Bragg condition for 222 reflection is satisfied.

Calculations of the structure factor in some crystals were given above. It is important to realize that the structure factor is independent of the shape and size of the unit cell. For example, two compounds of the type *AB* are given in Figure 5.15. Suppose that the unit cells of these two compounds differ in shape and size. If both contain two atoms per unit cell with fractional coordinates of A: (0,0,0) and B: (1/2,1/2,1/2), their structure factor is equivalently expressed by $F = f_A + f_B e^{\pi i(h+k+l)}$. As discussed earlier, the diffraction direction predicted by the Bragg law is determined solely by the shape and size of the unit cell. However, the intensities of diffracted beams are determined by the positions of atoms within the unit cell, being independent of its shape and size.

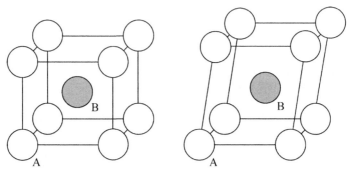

FIGURE 5.15 Two compounds of the type AB. Although the unit cells of these two compounds differ in shape and size, their structure factor is equivalently expressed by $F = f_A + f_B e^{\pi i(h+k+l)}$.

When the Bragg law is satisfied for an incident beam of intensity I_o, the diffraction intensity I_p from a small single crystal is predicted to be

$$I_p = I_e |F|^2 N^2 \qquad (5.26)$$

where

$$I_e = I_o \left(\frac{\mu_o e^2}{4\pi m R}\right)^2 \left(\frac{1+\cos^2 2\theta}{2}\right)$$

and N is the total number of unit cells in the crystal. It is to be noted that the intensity given by Eq. (5.26) is the maximum peak intensity obtainable only in ideal cases. As discussed in Section 4.5, the primary incident beam is neither perfectly parallel nor monochromatic. In addition, a crystal contains imperfections and is slightly mosaic in general. All these broaden the observed diffraction profile. Thus, the intensity value predicted by Eq. (5.26) is not an observable quantity. We already know that the diffrac-

Intensities of X-ray Diffraction 171

tion intensity is highest at the Bragg angle but still appreciable at angles slightly deviating from it. The *integrated intensity* of a diffraction peak is defined as the area under its I vs. 2θ curve. Meanwhile, the peak intensity refers to the maximum intensity observed with a diffraction peak, i.e., the intensity value at $2\theta_B$. Even for an infinite, perfect single crystal, the diffraction profile will be spread out by the various instrumental-broadening factors. While the peak intensity is a strong function of the instrumental factors, the integrated intensity is characteristic of the sample and is much less sensitive to them. In this respect, the integrated intensity is a more robust quantity than the peak intensity.

5.5 SYSTEMATIC ABSENCE

As graphically illustrated in Figure 4.10, the Bragg condition in reciprocal space is that in order for an incident X-ray beam to reflect from a set of (hkl) planes, the difference between the scattered and incident beam vectors should be equal to the reciprocal lattice vector $\mathbf{H}_{hkl}$. What happens if $F_{hkl} = 0$? Let's consider reflection from a set of (001) planes. If F_{001} is not zero, diffraction occurs when the Bragg condition is satisfied for this set of planes (Figure 5.16(a)). When $F_{001} = 0$, however, no diffraction will be observed although the Bragg condition is satisfied. This means that the corresponding reciprocal lattice point 001 does not exist, as illustrated in Figure 5.16(b). The reciprocal lattice points missing due to $F_{hkl} = 0$ are called *"systematic absences"*. Thus, the reciprocal lattice of a crystal structure should be constructed taking the systematic absence into account.

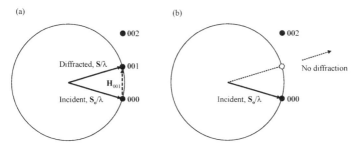

FIGURE 5.16 Bragg condition in reciprocal space vs. systematic absence. (a) If F_{001} is not zero, diffraction occurs when the Bragg condition is satisfied for a set of (001) planes. (b) When $F_{001} = 0$, the corresponding reciprocal lattice point 001 does not exist. Thus, diffraction signal is not observed even though the Bragg condition is satisfied.

When the Bragg law is satisfied for a set of (hkl) planes, the diffraction intensity from these planes is proportional to $\left|F_{hkl}\right|^2$. The structure factor may be real or a complex number. For a complex structure factor F, the squared magnitude can be obtained with multiplication by its complex conjugate F^*; $\left|F\right|^2 = FF^*$. The structure factors for hkl and $\overline{hkl}$ are complex conjugates to each other; i.e., $F_{hkl} = F^*_{\overline{hkl}}$. It means that the corresponding diffraction intensities are equal: $I_{hkl} = I_{\overline{hkl}}$. Thus, the diffraction pattern of a crystal is centro-symmetric irrespective of whether the crystal actually contains a center of symmetry or not. This is equivalent to saying that the reciprocal lattice is always centro-symmetric, because the symmetry information provided by the diffraction pattern is the symmetry of the reciprocal lattice. Accordingly, the existence of the center of symmetry in the crystal cannot be inferred from the existence of the same operation in the diffraction pattern. As a consequence, the crystal may have lower symmetry than the one displayed by the diffraction pattern. For example, GaAs lacks a center of symmetry, as manifest with the different etching rates on (111) and ($\overline{1}\overline{1}\overline{1}$). However, the reflections from these two oppositely-directed surfaces have equal diffraction intensities. For this reason, there will be some uncertainty when trying to determine the crystal symmetry from the diffraction experiment. Fortunately, some symmetry operations show their "footprints" in the reciprocal space and certain types of reflections from valid lattice planes produce no diffraction spots. This phenomenon is known as the systematic absence. The first type of systematic absence arises due to lattice centering. As a simple example, we consider a face-centered lattice of cubic symmetry. Many different structures such as NaCl, diamond, and zinc-blende structures have a face-centered cubic (FCC) lattice. Once the lattice type is face-centered, the structure factor is always zero for the lattice plane whose indices h, k, and l are mixed. This means that for a face-centered crystal, we do not expect to observe any intensity, for example, the (100), (210),... reflections. Consequently, the centering of a diffraction pattern that we obtain experimentally will tell us what particular type of lattice centering exists in real space. The other two symmetry elements, namely, glide planes and screw axes, also give rise to systematic absence. All these factors are already reflected in the structure factor, which depends on the arrangement of atoms within the unit cell.

It has been shown in Example 2.8 that the reciprocal lattice of an FCC structure with lattice constant "a" is BCC with lattice constant "$2/a$". In the earlier derivation of the reciprocal lattice, we started from the primitive

Intensities of X-ray Diffraction 173

cell vectors of the real lattice because the unit cell of the reciprocal lattice has been defined accordingly in Eq. (2.14). Now, the reciprocal lattice of a crystal structure can be derived straightforward from its conventional unit cell. The conventional cubic cell of an FCC structure with lattice constant "a" is represented by vectors **a**, **b**, and **c** in Figure 5.17. We neglect the face-centering translations of the structure and start from these conventional unit cell vectors. Then, three reciprocal vectors **a***, **b***, and **c*** obtained from Eq. (2.14) will be parallel to **a**, **b**, and **c**, respectively and have a length of $1/a$. Since F_{hkl} is zero for mixed hkl in the FCC structure, the reciprocal lattice points with corresponding indices (e.g., 100, 010, 001, 110, etc.) do not exist and should be removed from the as-derived lattice structure. This process leads to a BCC lattice with lattice constant "$2/a$", as depicted in Figure 5.17. It can be easily shown that the reciprocal lattice of a BCC structure with lattice constant "a" is FCC with lattice constant "$2/a$". For a simple cubic structure, there is no systematic absence because the structure factor $F_{hkl} = f$ is constant for any hkl reflections. Thus, the reciprocal lattice of a simple cubic structure is also simple cubic with lattice parameter of $1/a$. An example problem given below may help to be more familiar with the concept of systematic absence.

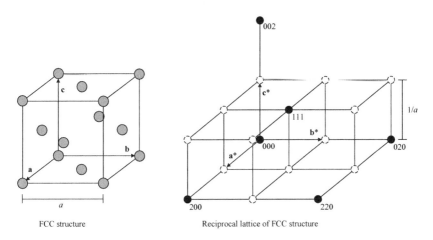

FCC structure   Reciprocal lattice of FCC structure

FIGURE 5.17 FCC structure and its reciprocal lattice.

174   X-Ray Diffraction for Materials Research: From Fundamentals to Applications

---

EXAMPLE 5.3

As discussed in Section 4.6, electron diffraction pattern directly visualizes the two-dimensional reciprocal lattice of a crystal. Let's suppose that an electron beam is incident along the $[\bar{1}\bar{1}0]$ direction of a thin (110)-oriented sample of FCC structure. Then,

(a) Draw the expected diffraction pattern and give an index to each spot.
(b) If the sample had a diamond structure, with all other conditions remaining unchanged, how would the diffraction pattern be modified?

Answer:

(a) As shown in Figure 5.18(a), the reciprocal lattice of FCC structure does not have lattice points for mixed Miller indices. Since the electron beam is incident along $[\bar{1}\bar{1}0]$, diffraction occurs by the reciprocal lattice vectors perpendicular to $[\bar{1}\bar{1}0]$. Some of these vectors are represented with bold arrows in Figure 5.18(a). When drawn from the origin, the reciprocal lattice vector $\mathbf{H}_{hkl}$ should satisfy the condition of $h + k = 0$, where h and k are unmixed. Then we will have a diffraction pattern like Figure 5.18(b). The reciprocal lattice of the FCC structure is BCC and the incident direction of the electron beam is $[\bar{1}\bar{1}0]$. Diffraction occurs from the reciprocal lattice points on a plane perpendicular to the incident beam direction. In cubic systems, $[\bar{1}\bar{1}0]$ is perpendicular to (110). Thus, the obtained electron diffraction pattern is consistent with the lattice point configuration on (110) of a BCC lattice. The size of diffraction spots decreases with increasing distance from the pattern center because the reciprocal lattice points (actually rods) are more unlikely to touch the Ewald sphere. In usual, the central spot is much stronger than the others and is often blocked with a long bar when taking a photo.

Intensities of X-ray Diffraction 175

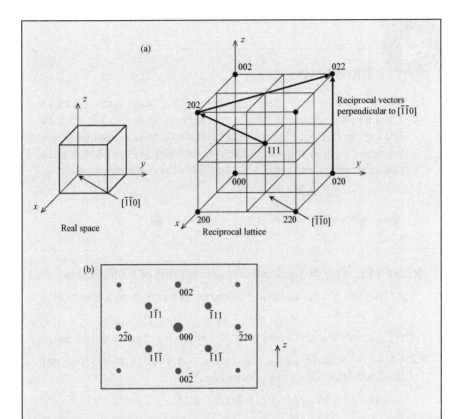

FIGURE 5.18 Diffraction of an electron beam incident along the [$\bar{1}\bar{1}0$] direction of a thin (110)-oriented FCC sample. (a) BCC reciprocal lattice of the FCC structure. (b) Electron diffraction pattern.

(b) In the diamond structure, the structure factor is also zero when ($h + k + l$) is an odd multiple of 2. So there are extra missing reciprocal lattice points in addition to those due to the mixed indices. Therefore, such spots as 002 and 00$\bar{2}$ in Figure 5.18(b) would not be observed in the electron diffraction pattern from a sample of diamond structure. Both of the FCC and diamond structures have an FCC lattice. However, their reciprocal lattices are different. While the FCC structure has a BCC reciprocal lattice, the reciprocal lattice of the diamond structure is not BCC anymore.

PROBLEMS

5.1. We have a row of N identical atoms with atomic scattering factor f and equally separated from each other, as shown in Figure 5.19. We expose this one-dimensional crystal to a monochromatic X-ray beam and wish to know the scattering intensity at P. When the phase difference between waves scattered from two adjacent atoms is δ,

FIGURE 5.19 One-dimensional crystal consisting of N identical atoms.

(a) Give the X-ray scattering intensity at point P, as a function of $N, f,$ and δ.

(b) Calculate the scattering intensity when we remove the mth atom.

5.2. AB compound $(a \neq b \neq c, \alpha = \beta = \gamma = 90^\circ)$ has 4 atoms per unit cell with the following coordinates.

A: (0,0,0), (1/2,1/2,0) B: (1/2,0,0), (0,1/2,0)

(a) Derive simplified expressions for $|F|^2$.

(b) What is the Bravais lattice of this material?

5.3. A material of simple tetragonal Bravais lattice (a = 2.4 Å, c = 3.6 Å) is prepared in a plate shape so that the sample surface is parallel to (001). When the sample was symmetrically scanned in the 2θ range from 20° to 100° using an X-ray beam at 1.54 Å, at which 2θ positions will the diffraction peaks be observed? How will the diffraction pattern change if the sample is replaced by a crystal of body-centered tetragonal lattice? Assume that all the other conditions are the same.

5.4. We have two different cubic crystals with the same lattice constant of a = 4.0 Å. One has the diamond structure and the other, zinc-blende structure. Both are made in plate shape and (001)-oriented. If we carry out diffraction experiments with λ = 1.54 Å and 2θ =

Intensities of X-ray Diffraction

20°–120°, will you find any difference between these two samples?

5.5. Three thin single-crystalline samples are put together as shown in Figure 5.20, and their diffraction patterns are measured. All samples have the same lattice parameter of $a = 3.6$ Å and their structures and orientations are given below. When the 2θ value changes from 20° to 120°, state the peak positions observed in each sample and from which planes the observed peaks come.

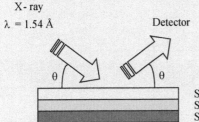

Sample A: (001)-oriented S.C structure
Sample B: (111)-oriented BCC structure
Sample C: (110)-oriented FCC structure

FIGURE 5.20 Diffraction from stacked thin single-crystalline samples.

5.6. Electron diffraction pattern directly represents the 2-D reciprocal lattice pattern. Let's consider the electron beam is incident along the [001] direction of a thin (001)-oriented Si sample. Draw the expected diffraction pattern and give an index to each spot (Hint: Consider the structure factor of Si. The reciprocal lattice vector corresponding to each spot is perpendicular to the incident direction of the electron).

5.7. The wurtzite form of ZnS is hexagonal with two Zn atoms and two S atoms per unit cell at positions:

Zn: (0,0,0), (1/3,2/3,1/2) S: (0,0,3/8), (1/3,2/3,7/8)

Derive simplified expressions for the structure factor F. For what hkl combinations will F vanish?

PART III
APPLICATIONS OF X-RAY DIFFRACTION

CHAPTER 6

CHARACTERIZATION OF THIN FILMS BY X-RAY DIFFRACTION

CONTENTS

6.1	Introduction	182
6.2	Four-Circle Diffractometer	186
6.3	Determination of In-Plane Orientations	191
6.4	Stress and Strain in Thin Films	203
6.5	Film Quality and Rocking Curve	210
6.6	Grazing Incidence X-Ray Diffraction	216
Problems		220

182   X-Ray Diffraction for Materials Research: From Fundamentals to Applications

6.1   INTRODUCTION

X-ray diffraction is a *powerful nondestructive technique* that can be utilized for phase identification, orientation determination, lattice parameter measurement, assessment of crystal quality, and determination of crystal structure. As discussed in the previous chapters, diffraction peaks are produced by constructive interference of X-rays scattered from a specific set of lattice planes. Since the peak intensities are determined by the atomic arrangement within the unit cell, the X-ray diffraction pattern is characteristic of a particular phase and material, providing a kind of fingerprint for comparison. Therefore, a large variety of crystalline samples can be quickly identified by a search of the standard database of X-ray diffraction patterns. In the Chapters 4 and 5, we have described the fundamental theory of X-ray diffraction. The common application areas of X-ray diffraction are discussed throughout the following three chapters, and we begin with the characterization of thin films. Modern electronic and opto-electronic devices are mostly made with thin films. Thin film analysis with X-rays requires no special sample preparation. It is a quick, non-contact method that can be used to determine important material parameters and predict device performance.

A *thin film* is a layer of material, which is typically deposited onto a substrate or previously deposited layers. Although "thin" is a relative term, a thin film usually refers to a layer with thickness less than 1 µm. Layers thicker than 1 µm are often called *thick films*. Thin films are widely used for optics and electronics. A familiar optics application is the household mirror, which has a thin metal layer coated on the back of a sheet of glass. Optical applications include reflective and anti-reflective coatings that consist of multiple layers having varying thicknesses and refractive indices. Most electronic devices require many different thin films that act as insulators, semiconductors, and conductors to form integrated circuits. Thin films are also useful to protect materials against corrosion, oxidation, and wear. Thin film deposition techniques are categorized into two primary methods, depending on whether the process is chemical or physical. The chemical method includes chemical vapor deposition, chemical solution deposition, and atomic layer deposition. Thermal evaporation, electron-beam evaporation, sputtering, and pulsed laser deposition are examples of the physical methods. Most deposition techniques can control layer thickness at the nanometer scales and some enable a single layer of atoms to be deposited at a time. For optimal performance, thin films are required to possess specific electrical, optical, and mechanical properties that strongly depend on their microstructure and crystallinity. The microstructure and crystallinity of films are significantly influenced by the substrate material and deposition conditions. Therefore, thin film characterization is essential to improve device quality to the acceptable level. X-ray diffraction is a very powerful nondestructive technique for characterizing thin films. It can provide a variety of information such as phase, lattice parameter, film thickness, orientation, internal stress and strain, etc. The purpose of this chapter is to introduce X-ray diffraction techniques that are commonly used to characterize thin films deposited on substrates. Before proceeding into the diffraction methods in detail, it is necessary to know some terminology relevant to thin film characterization.

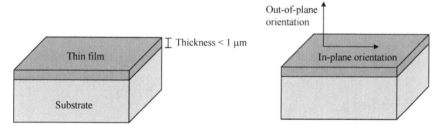

FIGURE 6.1 Definition of out-of-orientation and in-plane orientation in a thin film.

The *out-of-plane orientation* of film or substrate refers to the direction perpendicular to its surface (Figure 6.1). When a surface is known, its normal can be easily recognized. Therefore, it is more convenient to represent the out-of-plane orientations of the film and substrate by planes rather than directions. Thin films are mostly grown on single-crystal substrates, especially for electronic and optoelectronic applications. In these cases, the substrate surface is specified with Miller indices: e.g., (111)-oriented MgO and (001)-Si. Thus, if a film with (hkl) surface is deposited on a (001)-oriented substrate, the relationship between their out-of-plane orientations is simply expressed as $(hkl)_F // (001)_S$. Meanwhile, the *in-plane orientation* represents directions within the film and is parallel to the surface. Since it can be arbitrarily chosen, the in-plane orientation of a film relative to that of the substrate is stated by the angle between specific in-plane directions in two materials. Figure 6.2 shows an example where [100] and [010] of the film are at an angle of 45° to [100] of the (001)-oriented substrate. The relationship between their in-plane orientations is thus represented by $[010]_F//[110]_S$.

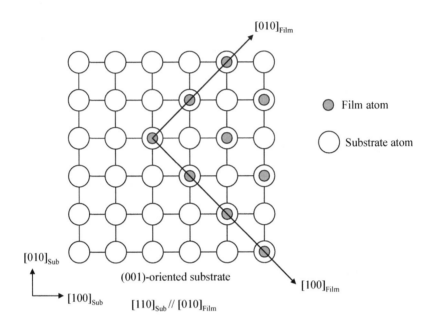

FIGURE 6.2 In-plane orientation relationship between the substrate and film.

A number of factors influence the microstructure of a thin film. Though deposited on a highly lattice-matched, single-crystalline substrate, the obtained film may be polycrystalline or even amorphous depending on the deposition conditions such as substrate temperature, gas pressure, and incident angle of the vapor beam. In general, the microstructures of crystalline films fall into three different categories, depending on the numbers of out-of-plane and in-plane orientations. When the film is perfectly single-crystalline, it has a single out-of-plane orientation and a single in-plane orientation, as depicted in Figure 6.3(a). A single-crystalline film is rarely achieved unless it is grown by homoepitaxy or deposited on a well lattice-matched substrate. Many thin films exhibit a single out-of-plane orientation and multiple in-plane orientations (Figure 6.3(b)), or multiple out-of-plane orientations (Figure 6.3(c)). In both cases, the films consist of many

Characterization of Thin Films by X-ray Diffraction

grains and are thus polycrystalline. In the former, all grains have the same out-of-plane orientation and only the in-plane orientations change across the grain boundary. On the contrary, the latter has multiple out-of-plane orientations and the surface of each grain has different Miller indices. The structures shown in Figures 6.3(b) and 6.3(c) represent two extreme states of a polycrystalline film. Actual films may have an intermediate state. Texture refers to the distribution of crystallographic orientations of a polycrystalline sample. A sample in which these orientations are completely random is said to have no texture. If the crystallographic orientations are not random, but have some preferred orientation, then the sample has a weak, moderate or strong texture. The degree of texture depends on the percentage of crystallites or grains having the preferred orientation. Texture can have a significant influence on materials properties. The structure given in Figure 6.3(b) is a completely textured structure, since all grains have the same out-of-plane orientations. When most of the grains have a specific orientation, with only a few having others, the film is said to be highly textured.

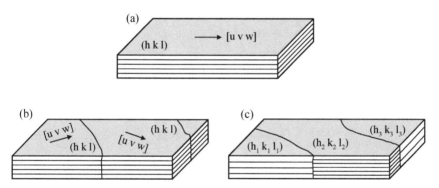

FIGURE 6.3 Three different types of thin films. (a) Single-crystalline film. (b) Polycrystalline film with a single out-of-plane orientation and multiple in-plane orientations. (c) Polycrystalline film with multiple out-of-plane orientations.

6.2 FOUR-CIRCLE DIFFRACTOMETER

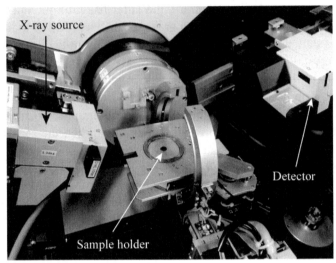

FIGURE 6.4 Four-circle diffractometer.

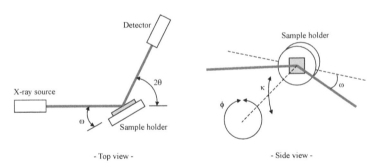

FIGURE 6.5 Configuration of four-circle diffractometer. 2θ, ω, κ, and ϕ angles can be varied independently.

Four-circle diffractometer is an X-ray diffraction apparatus that is widely utilized for characterizing thin films (Figure 6.4). Although the diffractometer may have different geometries and specifications depending on the manufacturer, its configurations are basically the same; the sample is stationed on a multi-axis goniometer that allows it to be rotated to a precise angular position. Figure 6.5 schematically illustrates the top and side views of the four-circle diffractometer. We here need to define the angles that appear in thin film characterization. 2θ means the angle between the incident and diffracted X-ray beams. ω refers to the angle between the incident X-ray beam and the sample holder. The sample holder can be tilted up and down and κ represents the tilting angle. The sample holder is also made to rotate around its surface normal and ϕ means this rotation angle. As described, the diffractometer consists of four circles allowing the attached sample to be brought into various orientations. Two circles, denoted by κ and ϕ, are used to adjust the crystal orientation relative to the diffractometer coordinate system. A third circle, ω, permits the orientation of the crystal lattice planes at a specific angle to the incident X-ray beam. Finally, the fourth circle, denoted by 2θ, moves the detector to lie at an angle of 2θ to the primary X-ray beam. The directions of the primary and diffracted beams should be in a horizontal plane so that each reciprocal lattice vector **H** can be brought into this diffraction plane. As illustrated in Figure 4.9, the incident beam vector, scattered beam vector, and reciprocal lattice vector should be coplanar in order for diffraction to occur. The four-circle diffractometer allows a specific reciprocal lattice vector of the sample to meet this condition by adjusting the ω, κ, and ϕ angles.

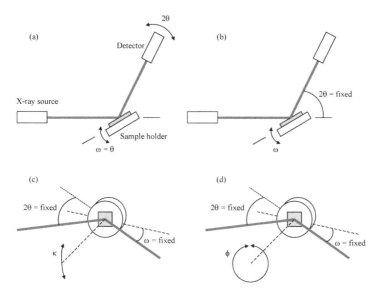

FIGURE 6.6 Four scanning modes: (a) θ-2θ scan, (b) ω scan, (c) κ scan, and (d) ϕ scan.

The diffractometer can be operated in four different scanning modes. In θ-2θ scan, both of the detector and sample holder are rotated while maintaining ω = θ (Figure 6.6(a)). This is a symmetric scan that enables the incident X-ray beam to be reflected from crystal planes parallel to the sample surface. In ω scan, the detector is fixed and only the sample holder is rotated as illustrated in Figure 6.6(b). This ω scan is useful to estimate the orientation spread of a film, i.e., the degree of mosaicity. In κ scan, the sample holder is tilted up and down with 2θ and ω fixed at certain values (Figure 6.6(c)). Here, κ = 0 represents the state where the normal to the sample holder becomes coplanar with the incident and diffracted directions. In φ scan, the planar sample holder is rotated 360° around its surface normal with all the other circles fixed, as depicted in Figure 6.6(d).

The out-of-plane orientations of a thin film are generally determined by four steps. Each step is here explained with a hypothetical sample. Figure 6.7(a) shows a thin film with (*hkl*) surface deposited on a (001)-oriented substrate. (*hkl*) of the film and (001) of the substrate are assumed to have non-zero structure factors.

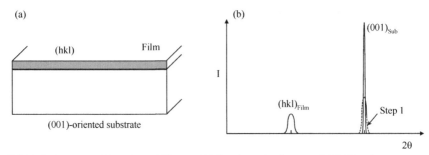

FIGURE 6.7 (a) (*hkl*)-oriented film on (001)-oriented substrate. (b) Diffraction patterns expected from the initial θ-2θ scan (step 1) and the final scan.

Step 1: First, θ-2θ scan is carried out for the substrate to find a 2θ value giving rise to the maximum peak intensity. This value is $2θ_{001}$ for the substrate.

Step 2: After fixing the detector at $2θ_{001}$, ω scan is performed to find out a ω value giving the maximum intensity.

Step 3: With the 2θ and ω values fixed as above, κ scan is carried out to determine a κ value giving the maximum intensity.

Step 4: Finally, θ-2θ scan is conducted again.

Thin film has much weaker diffraction peaks than the substrate. Then, a slight misalignment of the sample may cause no peaks from the film. The ω and κ angles are the instrumental angles measured with respect to the sample holder, not from the actual crystal planes. The substrate, usually cut from a single-crystal boule, may have a slightly different orientation from the designated one. The steps 2 and 3 are needed to adjust the crystal planes to the desired directions. As the film is constrained by the substrate, with its surface always parallel to the substrate surface, the substrate planes should be well aligned at first. If the substrate were cut to have a slightly different orientation from the intended (001) orientation, the corresponding reciprocal lattice vector $\mathbf{H}_{001}$ would not exactly bisect the angle between the X-ray source, sample holder, and detector when the detector was at $2\theta_{001}$ and the sample holder, at $\omega = \theta_{001}$. If this is the case, it is highly probable that only a substrate peak will be obtained in the step 1, as shown in Figure 6.7(b). The $2\theta_{001}$ value is determined by the interplanar spacing of the (001) planes and is independent of the orientation. If the substrate peak is rather weak and no film peaks are observed in the first θ-2θ scan, the Steps 2 and 3 are necessary. These two steps are tuning processes to maximize the diffraction intensity. If the substrate surface were not perfectly (001)-oriented, the ω and κ values giving rise to the maximum peak intensity would also slightly deviate from θ_{001} and 0, respectively. While the diffraction intensity may significantly change with ω, it is more or less insensitive to the κ value. The deviation angles are determined through the ω and κ scans, and the final θ-2θ scan is performed with these angles as offset angles. Then, we may have a stronger peak from the substrate than before, with the appearance of a film peak. When the substrate is simple cubic and the film has a body-centered cubic structure with its surface parallel to (110), the observed diffraction pattern will look like Figure 6.8. The structure factor of a body-centered structure is non-zero when $h + k + l$ are even. Thus, (220), (330), ..., peaks will also be obtained from the film. Since the substrate is simple cubic with no systematic absence, it will exhibit (001), (002),..., peaks. If the surface of the film is (001)-oriented, (002), (004), (006),..., peaks will be observed from the film. However, such peaks as (001), (003), and (005) are missing because the structure factor is zero for $h + k + l =$ odd.

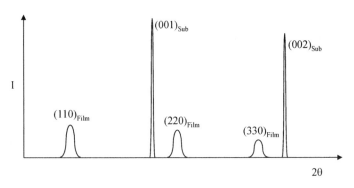

FIGURE 6.8 Diffraction pattern expected when a (110)-oriented film is deposited.

EXAMPLE 6.1

A thin film of BCC structure (a = 5.0 Å) was deposited on a (111)-oriented substrate of simple cubic structure (a = 4.0 Å). When the sample was symmetrically scanned in the 2θ range of 20–90° using a monochromatic X-ray beam of λ = 1.54 Å, a diffraction pattern shown in Figure 6.9 was obtained. Then, how many out-of-plane orientations exist in the film?

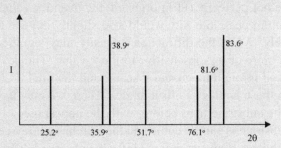

FIGURE 6.9 Diffraction pattern.

Answer: The spacing of a set of planes giving rise to a diffraction peak can be calculated from the Bragg law: $\lambda = 2d\sin\theta$. The interplanar spacings corresponding to the observed peaks are 3.530, 2.498, 2.310, 1.766, 1.249, 1.178, and 1.155 Å, respectively. Since the substrate is (111)-oriented, diffraction can occur from (111), (222), (333), ... planes. As it has d_{111} = 2.310 Å and d_{222} = 1.155 Å, the peaks observed at 2θ = 38.9° and 83.6° are from the substrate. The interplanar spacing of a cubic structure is given by $d_{hkl} = a/\sqrt{h^2+k^2+l^2}$. It is easily found that

Characterization of Thin Films by X-ray Diffraction 191

$3.530 \text{ Å} = a_{Film}/\sqrt{2}$

$2.498 \text{ Å} = a_{Film}/\sqrt{4}$

$1.755 \text{ Å} = a_{Film}/\sqrt{8}$

$1.249 \text{ Å} = a_{Film}/\sqrt{16}$

$1.178 \text{ Å} = a_{Film}/\sqrt{18}$

Thus, the peaks observed at $2\theta = 25.2°, 35.9°, 51.7°, 76.1°$, and $81.6°$ are from (110), (200), (220), (400), and (330) planes of the film, respectively. As $\sqrt{3^2 + 3^2 + 0^2} = \sqrt{4^2 + 1^2 + 1^1}$, the peak at $81.6°$ might have come from (411). However, it is more likely that this peak arises from (330) planes, because peaks from (110) and (220) planes that are parallel to (330) are observed together. Then, the film has grains with two different out-of-plane orientations: (100) and (110). Note that although some grains are (001)-oriented, (001) and (003) peaks are missing due to the systematic absence.

6.3 DETERMINATION OF IN-PLANE ORIENTATIONS

In the symmetric θ-2θ scan, reflection occurs from planes parallel to the sample surface. Thus, it provides information only on the out-of-plane orientation of a film. Figure 6.3(a) depicts the film that possesses a single out-of-plane orientation (hkl) and a single in-plane orientation. The film shown in Figure 6.3(b) exhibits the same out-of-plane orientation but has multiple in-plane orientations. The diffraction pattern of these two structures would be identical under the conventional θ-2θ scan, even though the former is perfectly single-crystalline and the latter is polycrystalline consisting of many grains. The symmetric scan detects only the out-of-plane orientation of a film and provides nothing about its in-plane orientation. Suppose that through the θ-2θ scan, we already know that the film in question has a single out-of-plane orientation. In order to find out whether it is truly single-crystalline or polycrystalline like Figure 6.3(b), we need to measure the in-plane orientation of the film, which can be revealed by the ϕ scan. No information on the in-plane orientation can be extracted from reflecting planes parallel to the film surface. Therefore, we should make the incident beam to be diffracted from planes suitably inclined to the surface. In the performed ϕ scan, the sample holder is first tilted and then rotated 360° around its normal with all the other circles fixed at certain values. This enables reflection to occur from a set of planes inclined to the sample surface. The in-plane orientation of a film is described relative to that of the substrate, by the angle between specific in-plane directions in two materials. Thus, the ϕ scan is separately performed for the substrate and the film. The resulting patterns are compared to derive their in-plane orientation relationship.

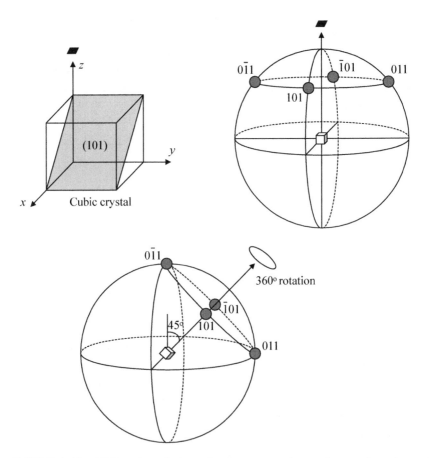

FIGURE 6.10 (101) and equivalent planes represented as poles on the sphere of projection in a cubic crystal.

Characterization of Thin Films by X-ray Diffraction 193

For the determination of in-plane orientations in a thin film, it is necessary to understand the geometry between planes in the film itself and their geometric relations with those of the substrate. As discussed in Section 2.7, the crystal planes are often represented as poles on the sphere of projection in order to figure out the associated symmetry and angular relationships more easily. When the normal of a plane is drawn from the sphere center to intersect the surface of the sphere, the intersecting point is called the pole of the plane. The orientation of a plane is represented by a pole on the sphere and the line connecting the sphere center to the pole is normal to the plane. Suppose a cubic crystal with 4-fold rotation symmetry along the z-axis is located at the center of a sphere, as shown in Figure 6.10. The poles of (101) and equivalent planes are marked as solid circles on the surface of the sphere. The line from the sphere center to any of the poles is at an angle of 45° to the 4-fold axis, i.e., the z-axis. When the crystal is tilted by 45° to any directions and then rotated 360° around the z-axis, the four poles alternately arrive at the north pole of the sphere every 90°.

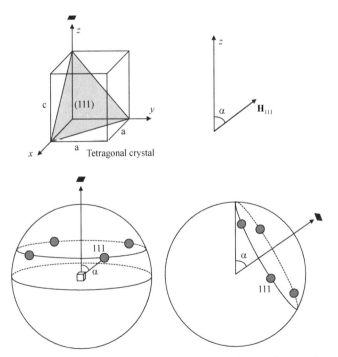

FIGURE 6.11 (111) and equivalent planes represented as poles on the sphere of projection in a tetragonal crystal.

Another example is given in Figure 6.11 for (111) and equivalent planes of a tetragonal crystal. The angle, α, between the 4-fold axis and the normal of {111} planes depends on the lattice parameters a, c of the crystal. This angle is equal to the angle between the z-axis and the reciprocal lattice vector $\mathbf{H}_{111}$. Similarly, if the crystal is tilted by α and then rotated around the z-axis, the {111} poles arrive at the north pole of the sphere every 90 degrees. The line connecting the sphere center to the (hkl) pole is parallel to the corresponding reciprocal lattice vector $\mathbf{H}_{hkl}$. Thus, when this crystal is rotated around its 4-fold axis after tilting, the four reciprocal lattice vectors take turns being vertically oriented every 90 degrees. In this case, the tilting angle is the angle between the 4-fold axis and the $\mathbf{H}_{hkl}$ vector.

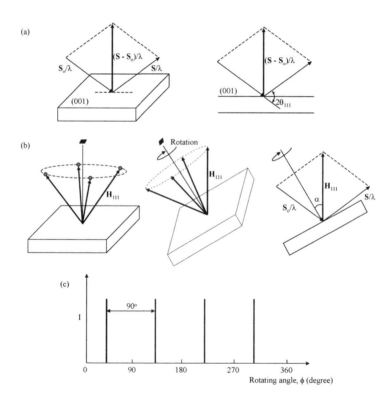

FIGURE 6.12 Diffraction from (111) plane in a (001)-oriented sample with four-fold rotational symmetry: (a) Diffraction condition for (111), (b) tilting and rotation of the sample, and (c) ϕ scan pattern.

Characterization of Thin Films by X-ray Diffraction

To find out the in-plane orientation of a two-dimensional material such as substrate and film, diffraction should be made to occur from a set of planes inclined to the surface because X-ray beams diffracted from planes parallel to it contain no information on the in-plane orientations. Of course, the plane selected for the in-plane orientation measurement must have a non-zero structure factor; planes of large structure factor are preferred. Suppose that a slab of crystal is (001)-oriented and has a 4-fold rotation axis perpendicular to the surface (see, Figure 6.12(a)). In order for diffraction to occur from {111} planes, the X-ray source and detector should first be arranged so that the angle between the incident and scattered beams becomes $2\theta_{111}$, which satisfies the Bragg law $\lambda = 2d_{111}\sin\theta_{111}$. Then, the difference between the incident and scattered beam vectors, $(\mathbf{S} - \mathbf{S}_o)/\lambda$, has a magnitude of $1/d_{111}$, which equals the length of $\mathbf{H}_{111}$. It is important to note here that the normal to (111) is not perpendicular to the sample surface. Although the reciprocal lattice vector $\mathbf{H}_{111}$ has the same magnitude as the $(\mathbf{S} - \mathbf{S}_o)/\lambda$ vector, diffraction does not occur because these two vectors are not parallel to each other. The four reciprocal lattice vectors are equally inclined to the symmetry axis, as shown in Figure 6.12(b) and the inclination angle depends on the lattice parameters of the crystal. Let this angle be α. If the crystal is tilted by α and subsequently rotated around the symmetry axis with the detector fixed at $2\theta_{111}$, the reciprocal lattice vectors alternately coincide with $(\mathbf{S} - \mathbf{S}_o)/\lambda$ every 90 degrees, satisfying the Bragg condition for diffraction. Then we have four peaks on 360° rotation and they are 90° separated from one another, as shown in Figure 6.12(c).

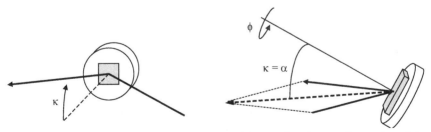

FIGURE 6.13 Sample can be tilted and rotated by varying the κ and φ angles of the sample holder.

The crystal can be tilted by changing the κ angle of the sample holder and rotated around the surface normal by varying its φ angle, as depicted in Figure 6.13. Here 2θ, ω, and κ are fixed at $2\theta_{111}$, θ_{111}, and α, respectively, and the diffraction intensity is measured as a function of ϕ. If the (111) peak is missing due to the systematic absence as in a BCC crystal, (222) can be used as the reflecting plane. In this case, the detector should be set at $2\theta_{222}$. Two factors should be taken into account when we choose the reflecting plane with which the ϕ scan is performed. First, it should have a structure factor as large as possible. Another consideration is that the plane to be analyzed needs to be suitably inclined to the surface, with the tilting angle of 30–60°. For the tilting angle beyond 60°, the sample holder may block the diffracted beams.

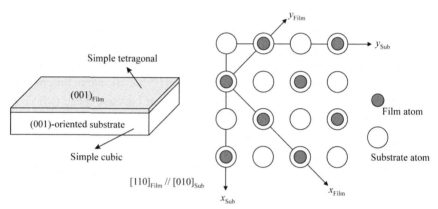

FIGURE 6.14 A single-crystalline film/substrate with an in-plane orientation relation of $[110]_{Film}/[010]_{Sub}$.

The in-plane orientation relationship between the film and substrate can be known by comparing the results of the ϕ scans that are separately conducted for the two materials. In actual analysis, the orientation relationship is deduced from the measured diffraction patterns. However, the underlying principle may be better explained in the opposite order, i.e., by deriving the expected diffraction patterns from a specific film/substrate combination. Figure 6.14 shows an example in which a single-crystalline film is deposited on a single-crystal substrate. The film has a simple tetragonal structure and the substrate, a simple cubic structure. Both of the substrate and film are (001)-oriented and their in-plane orientations have a relationship of $[110]_{Film}/[010]_{Sub}$. That is, the x-axis of the film is rotated by 45° from the x-axis of the substrate. As discussed before, a set of planes suitably inclined to the surface should be selected for the ϕ scan on this

Characterization of Thin Films by X-ray Diffraction

purpose. Here, (101) planes are selected for both the substrate and the film, as shown in Figure 6.15. Any set of planes has non-zero diffraction intensity in simple structures. Both of the substrate and film have four-fold rotational symmetry along the z-axis. Figure 6.15 also shows the poles of (101) and its equivalent planes marked on a sphere. Since the x-axis of the film is at 45° to the x-axis of the substrate, the {101} poles of the film are rotated by 45° around the z-axis with respect to those of the substrate. Therefore, the {101} diffraction peaks of the film are also shifted by 45° in ϕ from the substrate peaks. Although Figure 6.15 compares the {101} peaks of both materials in a single graph, the ϕ-scans are carried out separately because the required 2θ and κ values are different. The in-plane orientation relationship between the film and substrate can be deduced from the relative positions of the observed peaks. The absolute ϕ values do not have any meaning because they simply depend on how we attach the sample to the sample holder. Only the relative peak positions are meaningful. When (111) instead of (101) is selected for the ϕ scan of the film (Figure 6.16), the corresponding diffraction peaks will be observed at the same ϕ angles as the substrate {101} peaks. This is because the x-axis of the film is already rotated by 45° from that of the substrate and thus the {111} poles of the film have the same rotation angles as the {101} poles of the substrate.

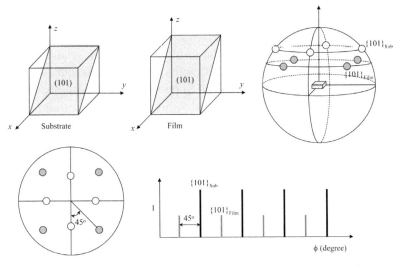

FIGURE 6.15 {101} poles of the substrate and film and their diffraction peaks.

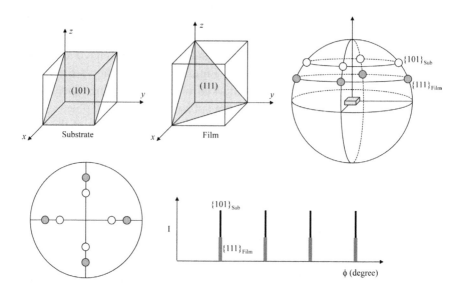

FIGURE 6.16 Substrate {101} and film {111} poles and their diffraction peaks.

As described above, any planes inclined to the surface can be chosen for the ϕ scans unless their reciprocal lattice points are systematically absent. To select a proper plane for the ϕ-scan, the out-of-plane orientation of the film should be known in advance so that the sample can be tilted in accordance with the Bragg condition for diffraction. Therefore, the ϕ scan is usually preceded by a symmetric θ-2θ scan to determine the out-of-plane orientation of the film. The in-plane orientation relationship can then be derived by comparing the diffraction patterns of ϕ-scans separately performed for the substrate and film. Suppose that the substrate and film are already known to be (001)-oriented through the θ-2θ scan, and thus we selected (101) planes for the ϕ scans in both materials. If the film {101} peaks were observed at angular positions shifted by β from those of the substrate {101} peaks, it means that $<100>_{Film}$ makes an angle of β with $<100>_{Sub}$ on the surface plane, as illustrated in Figure 6.17.

Characterization of Thin Films by X-ray Diffraction 199

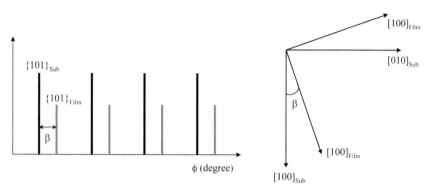

FIGURE 6.17 Angular separation of the peaks vs. in-plane orientation relation.

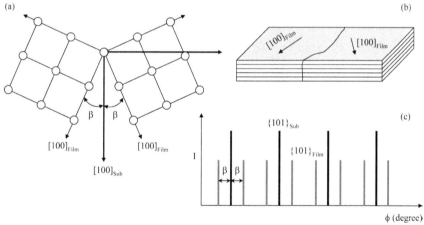

FIGURE 6.18 (a) In cubic and tetragonal systems, it is equally probable that the film is oriented with its [100] rotated either by $+\beta$ or $-\beta$ from [100] of the substrate. (b) Polycrystalline film with two in-plane orientations. (c) Diffraction pattern obtained when the in-plane orientation relation is like (a).

A practical aspect is worthy of consideration in this case. In the cubic and tetragonal structures, [100] and [010] directions are crystallographically identical. Thus, in the actual thin film deposition, it is equally probable that the film is oriented with its [100] rotated either by $+\beta$ or $-\beta$ from [100] of the substrate, as depicted in Figure 6.18(a). This will lead to a polycrystalline film with two in-plane orientations (Figure 6.18(b)). As a result, eight ϕ peaks will be observed from the film (Figure 6.18(c)). The appearance of eight peaks from a crystal with four-fold symmetry confirms the existence of two in-plane orientations. When β becomes 0° or 45°, a single-crystalline film can be obtainable and then the number of observed peaks renders to four, as depicted in Figure 6.19. A number of factors affect the orientation and crystalline quality of a thin film deposited on the rigid substrate. One of the most critical factors is the lattice match between the film and substrate. To grow a high-quality, single-crystalline thin film, a substrate of suitable structure and lattice parameter should be searched at first. Figure 6.20 is a schematic illustration of the well lattice-matched case. The ϕ scan is very useful to determine the in-plane orientation of a single-crystalline or highly textured film. However, there is no much need for determining the in-plane orientation in a film with multiple out-of-plane orientations like Figure 6.3(c). Moreover, the diffraction peaks (obtained from the ϕ scan) are very weak in this type of polycrystalline film.

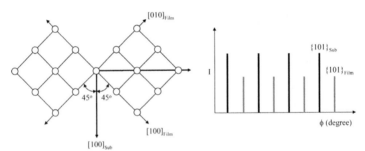

FIGURE 6.19 In-plane orientation relationship and the resulting diffraction pattern. When β is 45°, a single-crystalline film can be obtainable and four peaks are observed from the film.

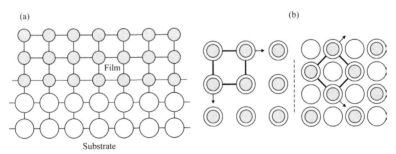

FIGURE 6.20 Well lattice-matched interfacial structure: (a) cross-section and (b) top view.

EXAMPLE 6.2

A thin film of FCC structure ($a = 4.60$ Å) was deposited on the substrate with an HCP structure ($a = 3.25$ Å) and their out-of-plane orientation was found to be $(111)_{Film}/(001)_{Sub}$. In this problem, the possibility that the film is single-crystalline is very high, when considering the interfacial structure and the lattice constants. But we like to confirm whether it is actually single-crystalline or not, through ϕ scans using $\lambda = 1.54$ Å.

(a) Select a plane of the substrate with which you perform the ϕ scan, and give 2θ and κ values required for this scan.
(b) Repeat (a) for the film.
(c) We finished scans both for the substrate and film. If the film is actually single-crystalline, how many peaks will you observe from each scan and what will their relative positions be?

Answer:

(a) When selecting a plane for the ϕ scan, two things must be taken into consideration. First, it should be inclined to the surface at some degrees. Second, its structure factor should not be zero. Since the structure factor of HCP is given by $F = f[1+e^{2\pi i(2h/3 + k/3 + l/2)}]$, (101), (102), ..., and so on can be selected. We here choose (102) as shown in Figure 6.21. The tilting angle κ is equal to the angle between the **c**-axis and $\mathbf{H}_{102}$. From the relation of $\tan\kappa = \dfrac{c/2}{\sqrt{3}a/2}$, κ is 44.33°. The 2θ value of 47.02° can be calculated from the interplanar spacing given by $d_{102} = \sqrt{3}a\sin\kappa/2$.

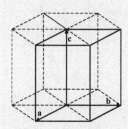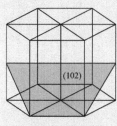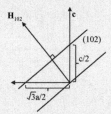

FIGURE 6.21 (102) plane in HCP structure.

(b) As the film is (111)-oriented, (200) is here selected for the ϕ-scan. It is to be noted that (100) cannot be chosen because the structure factor is zero. The tilting angle $\kappa = 54.74°$ is obtained by calculating the angle between the surface normal, i.e., [111] and $\mathbf{H}_{200}$. The Bragg law for (200) gives $2\theta_{200} = 39.12°$.

(c) The rotation axis of the substrate has six-fold rotation symmetry and thus a total of six peaks will be observed in the φ scan (Figure 6.22). The φ values of the peaks are determined by the angular positions of the reciprocal lattice vectors when they are projected onto a plane perpendicular to the rotation axis, that is, by the directions of the projected reciprocal lattice vectors on the surface plane. The rotation axis, [111], of the film has three-fold symmetry, giving rise to three peaks diffracted from the (200), (020), and (002) planes. The projected H_{200} and equivalent vectors have angular positions coincident with three of the six projected H_{102} vectors of the substrate. Therefore, the film {200} peaks will have the same φ angles as three of the substrate peaks.

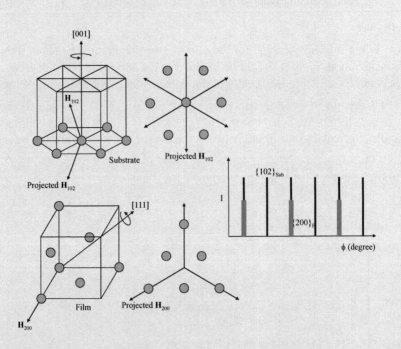

FIGURE 6.22 Projected reciprocal lattice vectors vs. φ peak positions.

6.4 STRESS AND STRAIN IN THIN FILMS

As a starting point in the discussion of stress and strain, consider a uniform cylindrical bar of cross-sectional area A that is subjected to an axial tensile force P (Figure 6.23). The normal stress, σ, is defined as force per unit area and expressed as $\sigma = P/A$. The tensile stress is represented with a positive value and the compressive stress, a negative value. The stress induces a strain that is defined as the change in length per unit length. The strain, ε, is then given by $\varepsilon = \Delta l/l_o$, where l_o is the original length and Δl, the length change. While the stress has units of N/m², the strain is a dimensionless quantity. In addition to the defined normal stress and strain, there are also shear stress and shear strain in mechanics. However, they are not directly measurable by X-ray diffraction, thus not being dealt with here. All solid materials can be deformed when subjected to an external load. It is already known from our daily experiences that up to a certain limiting load, a solid will recover its original dimensions when the load is removed. This is known as *elastic deformation or behavior*. The limiting load beyond which a material no longer behaves elastically is the elastic limit. When the elastic limit is exceeded, the loaded material has a permanently remnant deformation even after the load is removed.

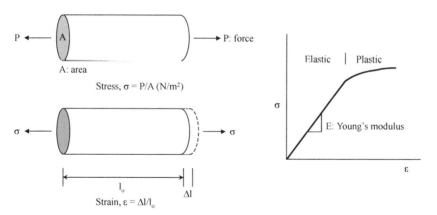

FIGURE 6.23 Definition of stress and strain, and a typical relation.

A material that is permanently deformed is said to have undergone *plastic deformation*. As long as the load is below the elastic limit, the deformation is proportional to the load in most solid materials. This relationship is known as Hooke's law, which governs the elastic behavior of a material. The load-deformation relationship is more frequently expressed as stress vs. strain. A typical σ-ε curve encountered in the solid material is also given in Figure 6.23. In the elastic region, the stress-strain relationship is linear with a slope of $E = \sigma/\varepsilon$, known as Young's modulus. E is therefore a measure of the stiffness of an elastic material and has units of pressure (N/m^2 or Pa). Young's modulus is not always the same in all orientations of a material. Here we shall only consider isotropic elastic solids. The theory of plasticity deals with the behavior of materials at strains where Hooke's law is no longer valid. A number of factors make the mathematical formulation of plasticity much more difficult than the description of the elastic behavior. For example, plastic strain is a function of the loading path by which the final state is reached, while the elastic deformation depends only on the initial and final states of stress and strain. The stress and strain in a thin film are largely caused by a lattice mismatch with the supporting substrate. The lattice mismatch-induced deformation is more likely to be elastic, rather than plastic.

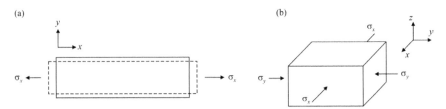

FIGURE 6.24 (a) Uniaxial and (b) biaxial stress systems.

While a tensile force applied in the x direction produces an elongation along that direction, it also causes a contraction in the transverse y and z directions (Figure 6.24(a)). It has been found that the transverse strains are a constant fraction of the strain in the longitudinal direction. These strains are related by the following equation.

$$\varepsilon_y = \varepsilon_z = -\nu\varepsilon_x = -\nu\frac{\sigma_x}{E} \qquad (6.1)$$

where ν is Poisson's ratio. The uniaxial stress σ_x produces a normal strain $\varepsilon_x = \sigma_x/E$ and two transverse strains $\varepsilon_y = -\nu\varepsilon_x$ and $\varepsilon_z = -\nu\varepsilon_x$. The values of ν are close to 1/3 for most metals. The strain produced by more

Characterization of Thin Films by X-ray Diffraction

than one stress component can be determined by applying the principle of superposition. The stress-strain relations for a three-dimensional state of stress are then given by

$$\varepsilon_x = \frac{1}{E}[\sigma_x - v(\sigma_y + \sigma_z)]$$

$$\varepsilon_y = \frac{1}{E}[\sigma_y - v(\sigma_z + \sigma_x)]$$

$$\varepsilon_z = \frac{1}{E}[\sigma_z - v(\sigma_x + \sigma_y)] \tag{6.2}$$

Many problems can be simplified for a two-dimensional state of stress like Figure 6.24(b), which consists of two normal stresses σ_x and σ_y. This biaxial stress system is frequently encountered when one of the dimensions of the body is small relative to the others. A typical example of the two-dimensional body is a thin film. In fact, there is no stress acting perpendicular to the free surface of the film. In a thin film, Eq. (6.2) reduces to

$$\varepsilon_x = \frac{1}{E}[\sigma_x - v\sigma_y]$$

$$\varepsilon_y = \frac{1}{E}[\sigma_y - v\sigma_x]$$

$$\varepsilon_z = -\frac{v}{E}(\sigma_x + \sigma_y) \tag{6.3}$$

$$\varepsilon_x + \varepsilon_y = \frac{1}{E}(1-v)(\sigma_x + \sigma_y)$$

$$\varepsilon_z = -\frac{v}{1-v}(\varepsilon_x + \varepsilon_y) \tag{6.4}$$

206 X-Ray Diffraction for Materials Research: From Fundamentals to Applications

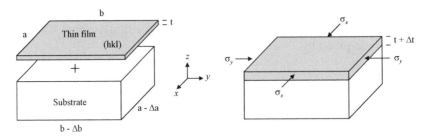

FIGURE 6.25 Schematic showing how a thin film got to have residual stress and strain.

How a thin film got to have residual stress and strain can be explained with a simple geometry given in Figure 6.25. There is a free-standing film with dimensions a and b along the x and y directions. Suppose that we like to attach it to a substrate with slightly smaller dimensions of $a-\Delta a$ and $b-\Delta b$ along the same directions. If this film needs to be side-matched with the substrate, it should first be contracted in dimensions, before being glued to the substrate. This requires that compressive stresses σ_x and σ_y be applied to the film. To keep the film being attached to the substrate with sides matched, it must be under the same stress fields. Otherwise, the film would be detached from the substrate and recover its original dimensions. In other words, once the film remains constrained by the substrate, it has residual stresses. The compressive stresses σ_x and σ_y residual in the film causes a thickness increase from t to $t + \Delta t$. Assuming that the film is (hkl)-oriented, the interplanar spacing would change from d_{hkl} to $d_{hkl} + \Delta d_{hkl}$. It will result in a shift of the (hkl) peak position from $2\theta_o$ to $2\theta_s$ (Figure 6.26). In this case, the strains within the film are given by

$$\varepsilon_x = \frac{1}{E}\left[\sigma_x - \nu\sigma_y\right] = -\Delta a / a$$

$$\varepsilon_y = \frac{1}{E}\left[\sigma_y - \nu\sigma_x\right] = -\Delta b / b$$

$$\varepsilon_z = -\frac{\nu}{1-\nu}\left(\varepsilon_x + \varepsilon_y\right) = \Delta t / t = \Delta d_{hkl} / d_{hkl} \qquad (6.5)$$

Characterization of Thin Films by X-ray Diffraction

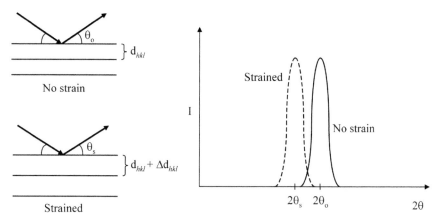

FIGURE 6.26 Peak shift arising from the change in interplanar spacing.

By differentiating the Bragg law: $\lambda = 2d \sin \theta$, we obtain the relation of $0 = 2 \sin \theta \Delta d + 2d \cos \theta \Delta \theta$. Manipulation of this relation gives

$$\Delta(2\theta) = 2\theta_s - 2\theta_o = -2 \tan \theta_o \varepsilon_z \tag{6.6}$$

where $2\theta_s$ is the experimentally measured position and $2\theta_o$, the theoretically calculated one. Eq. (6.6) makes it possible to determine a strain normal to the film surface by measuring a shift in the peak position.

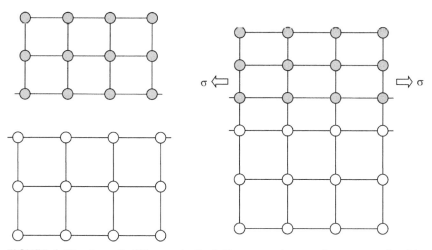

FIGURE 6.27 A small difference in the lattice parameters can be accommodated by strain energy. The misfit strain gives rise to an elastic stress in the film.

As mentioned previously, a major source of the elastic stresses developed in a thin film is the lattice mismatch between the film and substrate. These stresses arise when the films have coherent interfaces with their substrates. *Coherent interface* means an interface in which the crystal lattices (or atoms) match up on a 1-to-1 basis. Even in the case of perfect atomic matching, there is always a chemical contribution to the interface energy. *Incoherent interface* refers to an interface in which the atomic structure is disordered. The disordered structure results in a higher interfacial energy. Whether the film will exhibit a coherent or incoherent interface is determined by a competition between the elastic energy and the interfacial energy. A small difference in the lattice parameters, i.e., a small misfit, can be accommodated by strain energy and the misfit strain gives rise to an elastic stress in the film, as shown in Figure 6.27. The biaxial elastic stresses developed in a thin film depend on its lattice parameters relative to those of the substrate. If the film has smaller lattice parameters in both of the x and y directions, tensile stresses are generated along both directions (see Figure 6.28(a)). When it has a larger parameter along one direction and a smaller one along the other, the developed stresses will be mixed, as shown in Figure 6.28(b). In the latter case, two strain components induced along the z direction may cancel out each other. Then, the film will exhibit no strain normal to its surface. Different materials have different thermal expansion behaviors. Thus, when a film on a substrate is subjected to a temperature change, thermal stresses can be generated in the film and substrate. It is often observed that some stresses develop in films during deposition or growth. They are not due to lattice mismatch or thermal mismatch strains. These intrinsic stresses (or growth stresses) arise because thin films are usually deposited under non-equilibrium conditions. Since the film is constrained by the substrate, any redistribution of matter will result in stress. It is always strain that is directly measured by X-ray diffraction. The stress is determined indirectly, either by a calculation using such mechanics equations as Eqs. (6.2) and (6.3) or calibration. It is a challenge to exactly determine the residual stress of a thin film by X-ray diffraction alone, because its elastic constants (E and ν) are generally different from those at the bulk state. Nevertheless, X-ray diffraction is a useful tool for comparatively analyzing the residual stresses of thin films.

Characterization of Thin Films by X-ray Diffraction

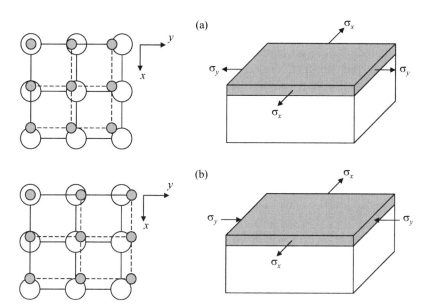

FIGURE 6.28 (a) When the film has smaller lattice parameters in both directions, tensile stresses are generated. (b) If it has a larger parameter along one direction and a smaller one along the other, the developed stresses are mixed.

EXAMPLE 6.3

Materials A and B have the same simple cubic structures. The lattice constant is generally a function of temperature, and those of A and B are given as below.

$$a_A = 4.00 + 2\times 10^{-4}T,\; a_B = 4.05 + 10^{-4}T$$

where T is the temperature in Kelvin and the lattice constant has units of Å. A thin film of material A was deposited on a (001)-oriented B substrate at $T = 500$ K. It was found that the film/substrate has a coherent interface with the orientation relations of $(001)_F//(001)_S$ and $[100]_F//[100]_S$. Then, calculate the elastic strains developed in the film when it was cooled to room temperature ($T = 298$ K). Also find the 2θ position of $(001)_F$ peak when θ-2θ scan is carried out with X-rays of $\lambda = 1.54$ Å. The Poisson's ratio of A is $\nu = 0.3$.

210   X-Ray Diffraction for Materials Research: From Fundamentals to Applications

> **Answer:**
>
> Both materials have an identical lattice parameter of 4.1 Å at 500 K. Therefore, the film and substrate are perfectly lattice-matched without any elastic strain. At 298 K, the materials A and B have lattice parameters of 4.06 Å and 4.08 Å, respectively. However, the thin film (material A) will exhibit the same in-plane lattice parameter as the substrate at room temperature, because it is supported and constrained by the substrate. This induces mutually orthogonal in-plane strains ε_x and ε_y in the film that have an equal magnitude of $\varepsilon_x = \varepsilon_y = 0.02 / 4.06 = 4.93 \times 10^{-3}$. From Eq. (6.4) and $v = 0.3$, we obtain $\varepsilon_z = -4.23 \times 10^{-3}$. Since tensile strains are developed in the film plane, a compressive strain is induced along the surface normal. As a result of the compressive strain, the film has $d_{001} = 4.04$ Å. Thus, the $(001)_F$ peak will be observed at $2\theta = 21.98°$. If there were no strain, it would have been obtained at $21.87°$.

6.5   FILM QUALITY AND ROCKING CURVE

The misfit that can be accommodated by elastic strain is limited. When a film constrained by a rigid substrate is very thin, it would be uniformly strained. Thus, the developed stress will also be uniform over the whole thickness, as illustrated in Figure 6.29. The elastic strain energy stored in a film increases in proportion to the film thickness. As the film gets thicker, top atomic layers would have its original lattice parameter to reduce the total energy. This will result in a nonuniform strain and stress (Figure 6.30). Namely, the top layers of the film are nearly free from strain and stress. Under this circumstance, the spacing of any particular set of planes varies with a distance from the interface. The nonuniform strain causes a broadening of the corresponding diffraction peak. In fact, the diffraction lines may be both shifted and broadened, because not only do the plane spacings vary from position to position but their mean value differs from that of the strain-free film. The mechanics equations of Eq. (6.3) to (6.6) have been derived on the assumption of uniform stress and strain fields. These expressions only hold when the film is sufficiently thin so that it can take the lattice parameter of the substrate. When the film thickness further increases and exceeds a critical value, it becomes energetically favorable for misfit dislocation at the interface to reduce the stress (Figure 6.31).

Characterization of Thin Films by X-ray Diffraction

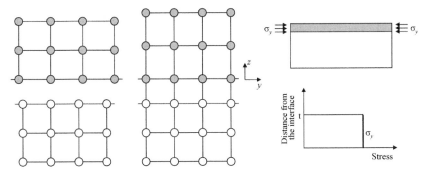

FIGURE 6.29 Uniform stress developed in a very thin film.

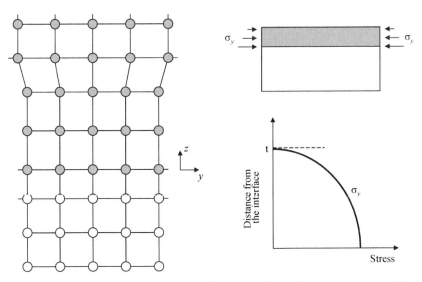

FIGURE 6.30 With increasing thickness, the developed stress becomes more nonuniform.

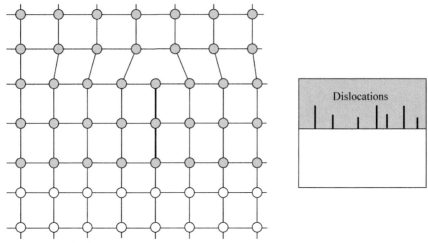

FIGURE 6.31 Misfit dislocation.

Dislocations are very common in epitaxially grown thin films and are spontaneously formed to relax the misfit strain. The appearance of misfit dislocations is influenced by the degree of lattice mismatch as well as the film thickness. If the lattice mismatch is fairly small, the misfit can be accommodated by the elastic strain energy and thus a dislocation-free epitaxial film may be grown up to a considerable thickness. On the contrary, dislocations will appear in a relatively thin film when the lattice mismatch becomes more profound. Figure 6.32 shows cross-sectional TEM images of a $Si_{0.7}Ge_{0.3}$ film grown on Si wafer. This epitaxial film is dislocation-free. Si and Ge have the same diamond structure. The lattice constants of Si and Ge are a_{Si} = 5.43 Å, a_{Ge} = 5.66 Å. Si_xGe_{1-x} has an intermediate lattice constant. When "x" is larger than 0.7, the lattice mismatch between Si_xGe_{1-x} and Si can be accommodated by the elastic strain formed in the Si_xGe_{1-x} film. The misfit increases with the decreasing value of "x". Ge and Si have a lattice misfit of 4.2%. Therefore, an epitaxial Ge film grown on Si substrate exhibits some dislocations, as shown in Figure 6.33(a). A high-resolution image of the interface clearly shows the existence of misfit dislocation (Figure 6.33(b)). The presence of dislocations may have an adverse effect on the electrical performance of semiconductor materials, providing easy diffusion pathways for dopants to cause short-circuits, or recombination centers to reduce carrier lifetime and density. The crys-

Characterization of Thin Films by X-ray Diffraction 213

talline quality of an epitaxial film is rapidly deteriorated with increasing density of dislocations.

FIGURE 6.32 Cross-sectional TEM images of a $Si_{0.7}Ge_{0.3}$ film grown on Si wafer (Courtesy: prof. D. Ko).

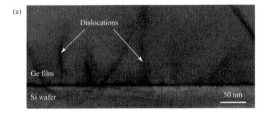

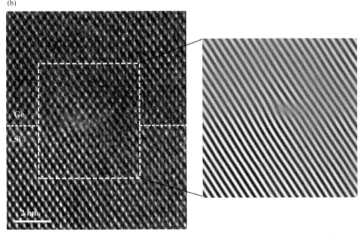

FIGURE 6.33 (a) Epitaxial Ge film grown on Si substrate. (b) High-resolution image of the Ge/Si interface (Courtesy: prof. D. Ko).

A *rocking curve* is a plot of Omega (ω) vs. X-ray intensity; it is obtained by changing the ω angle while keeping the X-ray source and detector stationary. From a rocking curve measurement, it is possible to determine the mean spread in orientation of the different crystalline domains of an imperfect crystal. When diffraction from a set of (*hkl*) planes parallel to the sample surface is concerned, the source and detector are fixed at $2\theta_{hkl}$ and only the sample is rotated (or "rocked"), as shown in Figure 6.34(a). Rocking curves are primarily used to study imperfections such as dislocations, mosaic spread, curvature, misorientation, and inhomogeneity. The rocking curve from a perfect crystal will have an intrinsic width arising from the instrument broadening and thickness effect. Different planes of a crystal also exhibit different intrinsic peak widths. Defects like mosaicity, dislocations, and curvature create disruptions in the perfect parallelism of atomic planes, causing the rocking curve to broaden beyond the intrinsic width for the Bragg peak. A crystal with mosaic structure does not have its atoms on a perfectly regular lattice throughout the whole crystal. Instead, the crystal is broken up into many tiny blocks, each slightly disoriented from one another. As a result, diffraction will occur not only at the Bragg angle but at other angles. The width of a rocking curve shown in Figure 6.34(b) is thus a combined product of the material and defects.

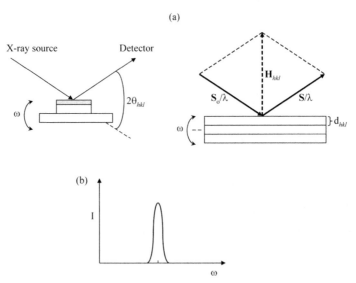

FIGURE 6.34 (a) Measurement of rocking curve. (b) Rocking curve has some width, which is a combined product of the material and defects.

Characterization of Thin Films by X-ray Diffraction 215

A thin film with high dislocation density has some angular spread in the crystal axes. This structure can be viewed as a structure consisting of mosaic blocks, each slightly disoriented from one another (Figure 6.35(a)). The angle of disorientation will increase with increasing dislocation density. The mosaic spread elongates the reciprocal lattice point in a direction parallel to the reflecting plane (*hkl*), causing the reciprocal lattice vector $\mathbf{H}_{hkl}$ to have some angular spread (Figure 6.35(b)). This will result in broadening of the rocking curve, because each mosaic block successively comes into diffraction position as the film is rotated (i.e., ω-scanned). The width of a rocking curve is a direct measure of the orientation distribution in a mosaic crystal. As the degree of mosaicity is related to the dislocation density in an epitaxial film without any curvature, it is possible to quantitatively analyze the dislocation density within a film by X-ray diffraction. The rocking curves are widely used for assessing the overall crystalline quality of thin films. It is to be noted that for the comparative analysis by rocking curves, the films should have similar thicknesses so that their intrinsic widths are not much different from one another.

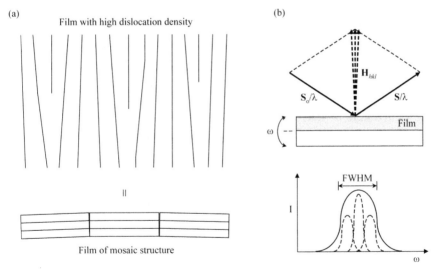

FIGURE 6.35 (a) Film with high dislocation density can be viewed as a structure consisting of mosaic blocks. (b) Angular spread of the reciprocal lattice vector in a mosaic structure and the resulting peak broadening.

6.6 GRAZING INCIDENCE X-RAY DIFFRACTION

X-ray radiation has a large penetration depth into any matter. Owing to this property, X-ray diffraction is not surface sensitive. Grazing incidence X-ray diffraction (GIXRD) is a technique to overcome this restriction. GIXRD measurements are performed at very low incident angles to maximize the signal from thin layers. It is sometimes very difficult to analyze thin films due to their low diffracted intensities, which results from small diffracting volumes compared to the substrates. When a thin film is characterized by the conventional θ-2θ scan (Figure 6.36(a)), diffraction signals from the film are much weaker than those from the substrate because the incident X-ray beam penetrates deeply into the substrate. Although the penetration depth varies with the symmetric sweep angle θ/2θ, it is generally much greater than the film thickness. The penetration depth of Cu K_α line at 2θ = 60° ranges from just above 1 μm for gold to 500 μm for graphite. When analyzing films much thinner than these values, strong scattering from the substrate may interfere with or completely drown out the weak signal from the film. The combination of low diffraction signal and high background makes it difficult to identify the phases present in the film. In the GIXRD geometry (Figure 6.36(b)), the stationary incident beam makes a very small angle with the sample surface (typically 0.3° to 3°), which increases the path length of the X-ray beam through the film. This can enhance the diffraction intensity of an ultrathin film, while considerably reducing the signal from the substrate at the same time. Since the path length is increased at grazing incidence, the diffracting volume of the film (i.e., its effective thickness) increases proportionally. As a result, there is a dramatic improvement in the film's signal-to-background ratio.

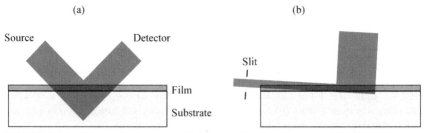

FIGURE 6.36 Schematic diagrams of (a) conventional θ-2θ scan and (b) GIXRD.

It is important to understand a fundamental difference between the conventional XRD based on a θ-2θ scan and the GIXRD. In the θ-2θ scan shown in Figure 6.37(a), the planes that contribute to diffraction peaks are always parallel to the surface. No information on any other planes can be obtained from this symmetric scan. With reference to the schematic geometry given in Figure 6.37(b), GIXRD detects planes that are tilted at an angle of θ – Φ from the surface, where Φ is the angle of incidence. In whatever geometry, the incident and diffracted beams are symmetric with respect to the reflecting planes. They make the same angle of θ with the reflecting planes. The incident angle Φ refers to the angle measured from the sample surface. In GIXRD, the incidence angle Φ is fixed and the angle (2θ) between the incident and diffracted beams is varied. When collecting the diffraction signal, only the detector is rotated through the angular range, keeping the incident angle, the beam path length, and the irradiated area constant. Under these conditions, crystal planes inclined to the sample surface are observed. The normal to these planes bisects the angle formed by the X-ray source, sample holder, and detector.

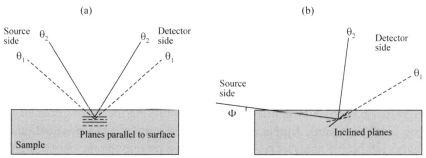

FIGURE 6.37 Schematic of the difference between (a) conventional XRD and (b) GIXRD.

The GIXRD method is not suitable, if the film has a high degree of preferred orientation (such as, an epitaxial film). In the symmetric θ-2θ scan, reflection occurs from planes that are parallel to the surface. Thus, an X-ray beam incident in any directions can be diffracted once the Bragg angle is maintained between the incident beam and the sample surface. It means that diffraction takes place independent of the incident direction. This is because the normal to the reflecting planes has no directional change even though the sample is rotated around its surface normal. When the reflecting planes are inclined to the surface, their normal varies in direction with the sample rotation. An epitaxial film is itself a single crystal, with a specific orientation relation with the underlying substrate material. Thus, diffraction at a particular wavelength only occurs when the sample is aligned correctly. For example, when the epitaxial film is $(h_1k_1l_1)$-oriented, a set of inclined $(h_2k_2l_2)$ planes would have a specific normal direction depending on the rotational position of the sample. In order for diffraction to occur from the $(h_2k_2l_2)$ planes, the X-ray beam should be incident in such a way that the normal to these planes becomes coplanar with a plane containing the X-ray source, sample, and detector all together.

The main purpose of the grazing incidence is to increase the effective thicknesses of very thin films, which are difficult to analyze by the typical θ-2θ scan. When the film is single-crystalline or highly textured, it should be accurately aligned with respect to the incident X-ray beam so that diffraction can occur from a particular set of planes inclined to the film surface. This requires information on the in-plane orientation of the film. However, there is no way of knowing it in advance. Therefore, GIXRD is not suitable for analyzing such thin films as shown in Figure 6.3(a) and (b). Meanwhile, a polycrystalline film like Figure 6.3(c) diffracts the incident X-ray beam in a conical fashion and some of the diffracted signals are captured by the detector. A polycrystalline film consists of a number of grains that have different orientations. As the normal to a set of (hkl) planes has random directions, the (hkl) planes in some grains will be properly oriented so that the diffraction condition is satisfied. The geometry of Figure 6.37(b) is primarily aimed for identifying the phases present in polycrystalline films. A great advantage of the grazing incidence method is that depth profile analysis is also possible by varying the incidence angle. In each scan, the angle of incidence is fixed so that the degree of penetration into the sample is kept constant throughout the measurement. At low incidence angles, the X-ray beam penetrates only the uppermost layers. At higher incidence angles, the X-rays penetrate deeper into the sample. Thus, successive layers can be sampled by adjusting the angle of grazing incidence.

Characterization of Thin Films by X-ray Diffraction 219

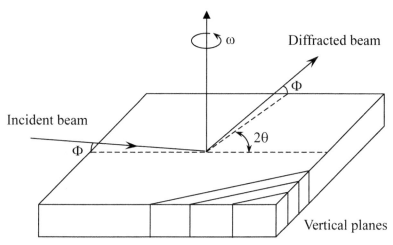

FIGURE 6.38 Grazing incidence in-plane X-ray diffraction (GIIXRD).

There is another configuration based on the grazing incidence, which is known as *grazing incidence in-plane X-ray diffraction* (GIIXRD). This geometry is illustrated in Figure 6.38. Both of the incident and diffracted beams are at grazing angles with respect to the sample surface. In GI-IXRD, the incident X-ray beam impinges onto the surface of a film at an angle of 1° or less, and the detector is placed in a horizontal plane nearly parallel to the film surface to detect diffraction from lattice planes that are perpendicular to the surface. GIIXRD data can be collected using a θ-2θ scan and/or an ω scan. In the θ-2θ scan, both of the film and detector are rotated, with the former rotating at half a rate which the detector is rotated. This is to record diffraction from planes perpendicular to a particular direction on the substrate. In the ω scan, the detector is stationary at a particular angle of 2θ and the film is rotated about its surface normal to record in-plane diffraction from a set of lattice planes that have a fixed value of spacing. In this respect, GIIXRD can be utilized to analyze the orientation relationship of an epitaxial film with the underlying substrate and measure the in-plane strains developed parallel to the film surface. Since the incident beam and the diffracted beam are both at very small angles with the sample surface, the use of Soller slits is necessary in both sides to ensure high angular resolution. This reduces the available X-ray intensity rather severely. Although GIIXRD experiments are also possible with laboratory X-ray sources, many such experiments are carried out by synchrotron radiation facilities that allow ultrathin films to be analyzed. In the symmetric θ-2θ diffraction, one-dimensional information along the surface normal is obtained. In GIIXRD, two-dimensional information on the surface is measured. The combination of both methods makes it possible to achieve three-dimensional information on any epitaxial films.

PROBLEMS

6.1. The following graph shows the diffraction pattern obtained with a θ-2θ scan on ZnO film/LiTaO$_3$ substrate. ZnO is hexagonal with a = 3.249 Å and c = 5.206 Å. LiTaO$_3$ is trigonal and its triple hexagonal unit cell has dimensions of a = 5.154 Å and c = 13.780 Å. The substrate is (100)-oriented. The Cu K_α line (λ = 1.54 Å) was used as the X-ray source. The position of the peaks are 2θ = 31.4° for ZnO and 62.0° for LiTaO$_3$. State which planes these peaks are coming from?

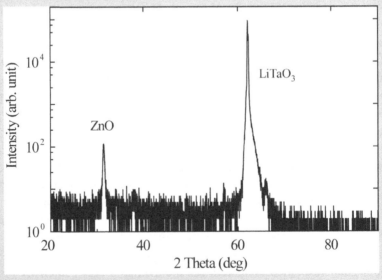

FIGURE 6.39 XRD pattern of ZnO film/LiTaO$_3$ substrate.

6.2. An AB_3 compound thin film (cubic with $a = 4$ Å) was deposited on a (001)-oriented substrate (body-centered tetragonal with $a = 4$ Å, $c = 5.6$ Å), with the out-of-plane orientation of $(001)_{Sub}//(001)_{Film}$. This compound has an order-disorder transition like $AuCu_3$. At $T < 100°C$, **A** atoms are located on the corners of the unit cell and **B** occupy the face-center positions. Above 100°C, all positions are randomly occupied by **A** and **B** atoms. We are carrying out θ-2θ scan for this sample in the scan range of 2θ = 30°–120° with an X-ray beam at λ = 1.54 Å.

(a) When the scan is performed at room temperature, identify all the observed peaks with their positions (2θ value).
(b) The same scan is carried out with the sample heated over the transition temperature. State if there are any changes to the result of (a). (Assume that there is no change in the lattice parameters).

6.3. MgO is cubic and $Sr_{0.5}Ba_{0.5}Nb_2O_6$ is tetragonal with 4mm point group. A $Sr_{0.5}Ba_{0.5}Nb_2O_6$ thin film was deposited on a (001)-oriented MgO substrate and their out-of-plane orientation relationship was found to be $(001)_S//(001)_F$. In order to find the in-plane orientation relationship between these two materials, ϕ scans were performed for the substrate (202) plane and the film (221) plane. The separately obtained two patterns are compared in the following graph. How many in-plane orientations exist in the film? What are the angles between $[100]_S$ and $[100]_F$?

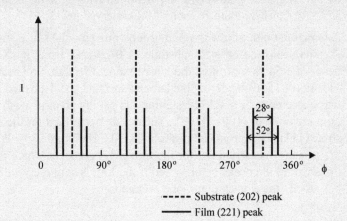

----- Substrate (202) peak
——— Film (221) peak

FIGURE 6.40 ϕ scan peaks.

222   X-Ray Diffraction for Materials Research: From Fundamentals to Applications

6.4. A (001)-oriented thin film (simple tetragonal structure, $a = 3$ Å, $c = 4$ Å) was deposited on the (001)-oriented substrate of simple cubic structure ($a = 3$ Å). When the sample is symmetrically scanned using a Cu K_α line ($\lambda = 1.54$ Å) in the range of $2\theta = 20°-120°$, draw the expected *diffraction intensity vs. 2θ* graph and state from which plane each of the peaks comes from.

6.5. θ-2θ scan ($\lambda = 1.54$ Å, $2\theta = 20°-120°$) for a Si film ($a = 5.43$ Å) deposited on the glass substrate revealed diffraction peaks at $2\theta = 28.43°, 47.31°, 56.38°$, and $106.82°$. How many out-of-plane orientations exist in the film?

6.6. In thin film growth, a buffer layer is often inserted between the film and the substrate when two materials have poor adhesion. Here is an example. After depositing a buffer layer **B** on the substrate **A**, a film **C** was deposited on top of the buffer layer. The substrate is simple cubic with $a = 2.4$ Å. The buffer layer has a body-centered tetragonal structure with $a = 2.4$ Å and $c = 3.4$ Å. The film possesses a simple orthogonal structure ($a = 3.0$ Å, $b = 2.4$ Å, and $c = 3.4$ Å). The out-of-plane orientation relationship is $(110)_A//(100)_B//(100)_C$. Explain about the diffraction pattern obtained with a symmetric scan performed in the 2θ range of $20°-100°$. $\lambda = 1.54$ Å.

6.7. A thin film of BCC structure ($a = 4.28$ Å) deposited on a Si substrate ($a = 5.43$ Å) revealed diffraction peaks at $2\theta = 42.18°, 52.29°$, and $69.08°$, when the θ-2θ scan was performed with an X-ray beam of $\lambda = 1.54$ Å. Then, identify each of the observed peaks.

6.8. A tetragonal thin film with point group 4mm ($a = 5.43$ Å, $c = 3.43$ Å) was deposited on a Si substrate. A θ-2θ scan using an X-ray beam of 1.54 Å confirmed the out-of-plane orientation relation of $(001)_S//(001)_F$. Considering the lattice constants, the deposited film was expected to be single-crystalline and exhibit an in-plane orientation relation of $[100]_S//[100]_F$. The ϕ scan performed for the substrate (111) plane produced peaks at $0°, 90°, 180°$, and $270°$. When the ϕ scan was carried out for the film (211) plane, at which positions would the peaks be observed if the film is single-crystalline, possessing the expected in-plane orientation.

Characterization of Thin Films by X-ray Diffraction 223

6.9. A thin film of BCC structure (a = 4.0 Å) was deposited on a (001)-oriented substrate of FCC structure (a = 5.65 Å). If the out-of-plane orientation relation is $(001)_S//(001)_F$, at which positions will the peaks be observed when a symmetric scan is carried out in the 2θ range of 20°–150° using a Cu K_α line (λ = 1.54 Å)?

6.10. A certain material having atomic radius of 1.5 Å can exhibit either an FCC or BCC structure in the thin film state. A thin film of this material showed diffraction peaks at 2θ = 52.85° and 125.50° under the θ-2θ scan performed using an X-ray beam at 1.54 Å. What are the structure and out-of-plane orientation of this film?

CHAPTER 7

LAUE METHOD AND DETERMINATION OF SINGLE CRYSTAL ORIENTATION

CONTENTS

7.1 Introduction 226

7.2 Laue Method 229

7.3 Indexing of Diffraction Spot 240

Problems 247

7.1 INTRODUCTION

The Laue method is mainly used to determine the orientation of single crystals. It reproduces von Laue's original experiment that was the first diffraction method ever used. The Bragg equation: $\lambda = 2d\sin\theta$ imposes very stringent conditions on λ and θ for any single crystal. When a parallel monochromatic X-ray beam falls on a stationary crystal, very few planes will be oriented so as to satisfy the Bragg law and as a result, very few reflections will be observed. To increase the number of reflections, either λ or θ should be continuously varied during the experiment. In the Laue method, a white X-rays beam (i.e., the continuous spectrum from an X-ray tube) is made to fall on a stationary single crystal. The Bragg angle θ is then fixed for every set of planes in the crystal. Each set of planes selects and diffracts a particular wavelength from the white radiation that satisfies the Bragg law for the values of d and θ involved. Thus, each diffracted beam has a different wavelength. The diffracted spots are recorded on a flat photographic film placed perpendicular to the incident X-ray beam. The symmetry of this Laue pattern corresponds to the symmetry of the crystal and directions of the crystal axes are determined by the symmetry axes of the Laue pattern. It allows a volume of as-grown single crystal to be cut in specific orientations. This begins with finding out the characteristic symmetry axes of the crystals: for example, the 4-fold rotational axis in tetragonal and <100> or <111> directions in cubic crystals.

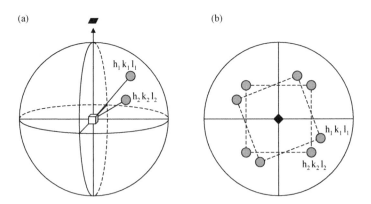

FIGURE 7.1 (a) Representation of crystal planes as poles on the projection sphere. (b) Stereographic projection of two crystal planes and their equivalent planes in a tetragonal crystal.

The underlying principle of Laue diffraction may be better explained with the projection of crystal planes. As discussed in Chapter 2, the symmetry of a crystal can be more easily figured out by representing the orientations of planes on a two-dimensional diagram. Suppose that a tetragonal crystal is positioned at the center of a projection sphere with its 4-fold rotational axis directed in the north, as shown in Figure 7.1(a). The arbitrary $(h_1k_1l_1)$ and $(h_2k_2l_2)$ planes marked as poles in Figure 7.1(a) are stereographically represented in Figure 7.1(b) along with their equivalent planes. This stereographic diagram clearly shows a 4-fold rotational symmetry axis perpendicular to the plane. This four-fold symmetry is still preserved with more projected poles from some other planes, since the crystal is tetragonal. In the stereographic projection, the plane normals are projected onto the equatorial plane of a projection sphere with its north and south poles as the reference points. For the understanding of Laue diffraction, it is more helpful to make use of the *gnomonic projection*, in which the point of projection is the sphere center.

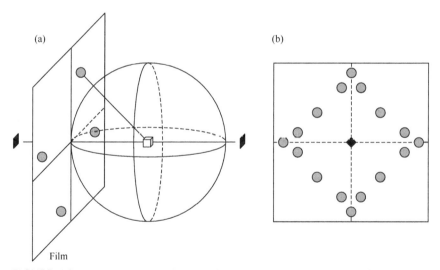

FIGURE 7.2 (a) Gnomonic projection of an arbitrary crystal plane and its equivalent planes. (b) Example of the gnomonic projection for many different planes in a tetragonal crystal.

In Figure 7.2(a), an arbitrary crystal plane and its equivalent planes are gnomonically projected onto a flat film vertically attached to the sphere of projection, where the four-fold axis of the crystal is set perpendicular to the film. Figure 7.2(b) shows an example of the gnomonic projection for many different planes. The projected poles exhibit a four-fold axis vertically running through the center of the film. When the characteristic symmetry axis of the crystal (here, the 4-fold axis) is inclined from the normal to the film surface, the symmetry center of the projected poles will be off the film center, as shown in Figure 7.3. If the Laue experiment in question is ultimately to cut the crystal normal or parallel to its characteristic symmetry axis, we need to tilt and/or rotate the crystal by adjusting the sample holder so that the symmetry center of the poles becomes coincident with the center of the film.

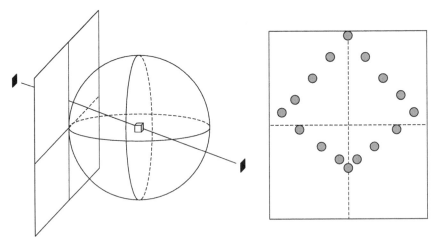

FIGURE 7.3 Gnomonically projected poles expected when the characteristic symmetry axis of the crystal is inclined from the normal to the film surface.

Suppose that an X-ray beam is incident parallel to the 4-fold axis of a tetragonal crystal through the center of a film, as illustrated in Figure 7.4(a). The plane normal always bisects the angle formed by the incident and diffracted beams. Therefore, the diffraction spots recorded on the film will have the same symmetry as the projected planes (Figure 7.4(b)). As-grown single crystals often have arbitrary shapes. Utilizing the diffraction patterns, they can be cut in specific orientations and prepared in the form of wafer, plate, cube, and others. Laue diffraction is also very useful to determine the longitudinal orientation of such one-dimensional crystals as wire and rod. Single crystals are usually anisotropic, with their physical and chemical properties depending on the orientation. In this respect, the crystals should be prepared in particular orientations not only for research purposes but also for device applications. This chapter describes the principle of Laue method, along with how the crystal orientation can be determined.

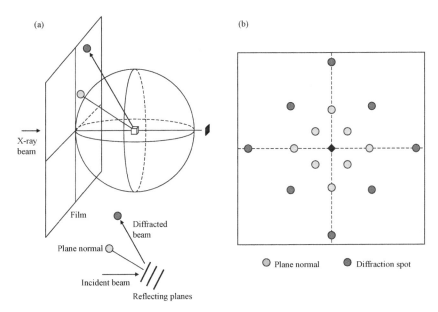

FIGURE 7.4 (a) Schematic illustration of the formation of a diffraction spot. (b) Symmetry equivalence of the projected plane normals and diffraction spots. It is to be noted that only the diffraction spots are recorded on the film.

7.2 LAUE METHOD

The Laue method is the oldest of the X-ray diffraction methods, in which a collimated beam of continuous spectrum falls on a stationary crystal. A broad X-ray spectrum in the incident beam is achieved by utilizing the unfiltered output from an X-ray tube. Since the intensity of the continuous spectrum increases with the atomic number of the target element, it is preferable to use a heavy-metal target, such as tungsten (W), but the unfiltered radiation from a copper target also does very well. For each set of (*hkl*) planes, the interplanar spacing d and the incident angle θ are fixed. A diffracted beam will be generated if the wavelength that satisfies the Bragg law is contained in the spectrum, as shown in Figure 7.5. For instance, if the incident white X-ray beam makes an angle of θ_1 with a set of ($h_1k_1l_1$) planes whose interplanar spacing is d_1, the λ_1 component will be diffracted in which the Bragg law of $\lambda_1 = 2d_1\sin\theta_1$ is applied. Similarly, a set of ($h_2k_2l_2$) planes with θ_2 and d_2 will diffract the λ_2 component in accordance with $\lambda_2 = 2d_2\sin\theta_2$. The different diffracted beams have different wavelengths and hence a Laue pattern is "colored". If X-rays were visible to the naked eyes like ordinary lights, each diffracted beam would exhibit a different color. Although these colors cannot be seen by our eyes, colored Laue patterns may be obtained by special photographic procedures. The positions of diffracted spots on the film depend on the crystal orientation relative to the incident beam. If the crystal is symmetrically oriented with respect to the primary beam, the resulting Laue pattern will also show such a symmetry since it directly reflects the symmetry of the crystal. If the crystal is bent or twisted, the diffraction spots become distorted and smeared out. In this respect, the Laue method is useful for both the determination of crystal orientation and the assessment of crystal quality.

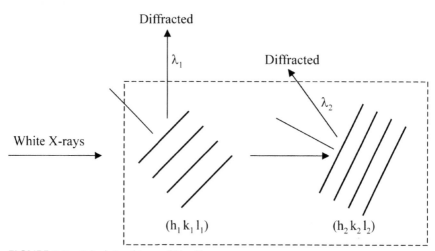

FIGURE 7.5 Principle of Laue diffraction.

The reciprocal-space treatment of Laue diffraction is given in Figure 7.6 for a particular set of (*hkl*) planes. As the incident beam has multiple wavelengths, it is represented by a series of parallel vectors, each with a different length equal to $1/\lambda$. All these vectors terminate at the origin of the reciprocal lattice. The origins of the incident beam vectors are the center of the Ewald sphere. Since the Ewald sphere now has varying radius, these vectors have different origins. Diffraction occurs when the reciprocal lattice vector $\mathbf{H}_{hkl}$ terminates on the surface of the Ewald sphere. The diffraction direction is given by a vector drawn from the sphere origin to the tip of $\mathbf{H}_{hkl}$. Of course, the wavelength of the diffracted beam is the reciprocal of the radius of this Ewald sphere. It is to be noted that higher-order reflections from $(2h\ 2k\ 2l)$ and $(3h\ 3k\ 3l)$ planes are also possible, if the incident white X-ray beam has a very wide spectral range. These reflections have the same Bragg angle as the (hkl) reflection and so all reflections will be superimposed on the same spot. When the wavelength of the (hkl) reflection is λ, the diffracted beams from the $(2h\ 2k\ 2l)$ and $(3h\ 3k\ 3l)$ planes have wavelengths of $\lambda/2$ and $\lambda/3$, respectively. Different spots have different wavelengths, but a few different wavelengths of integer multiples may also be mixed on the same spot. The spectral range of wavelengths in the incident beam is not infinitely wide but has lower and upper limits, which depend on the used X-ray tube and some other experimental factors.

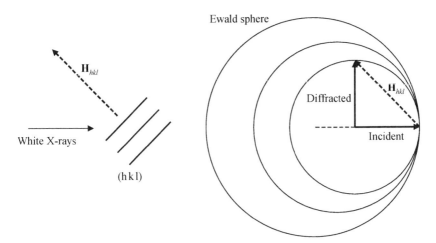

FIGURE 7.6 Laue diffraction condition for a set of (hkl) planes in reciprocal space.

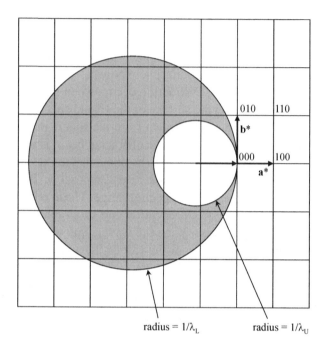

FIGURE 7.7 Reciprocal-space mapping of diffraction under a continuous X-ray spectrum of finite spectral width.

If λ_L and λ_U are the lower and upper wavelength limits, respectively, the largest Ewald sphere has a radius equal to $1/\lambda_L$ and the smallest one, a radius of $1/\lambda_U$ (Figure 7.7). There are a series of Ewald spheres between these two extremes. Any reciprocal lattice point lying in the shaded region of Figure 7.7 is on the surface of one of these Ewald spheres. It represents a set of crystal planes oriented to diffract one of the multiple wavelengths. As the spectral range of the incident beam gets broader, the number of reciprocal lattice points within the shaded region increases. Thus, more diffraction spots are obtained. Diffraction can take place either forward or backward. The direction of reflection from a particular set of planes can be easily found on the reciprocal diagram by drawing a circle passing through the origin of the reciprocal lattice and the corresponding lattice point. Then, a vector drawn from the center of this circle to the reciprocal lattice point represents the diffraction direction. For example, when the primary beam is incident along the $\mathbf{a}^*$ vector, i.e., perpendicular to the (h00) planes of the crystal, $\bar{3}10$ reflection occurs in the backward direction while $\bar{2}20$ reflection takes place in the forward direction (Figure 7.8). The real-space illustrations for these reflections are given in Figure 7.9. Laue method is the only X-ray diffraction technique that utilizes a white radiation consisting of many wavelength components. This is because the materials to be analyzed are single crystals. Each set of crystal planes has fixed spacing and orientation and thus selects a particular wavelength component that satisfies the Bragg law. Of course, there are no reflections from the planes whose structure factors are zero, even though the Bragg law may be satisfied.

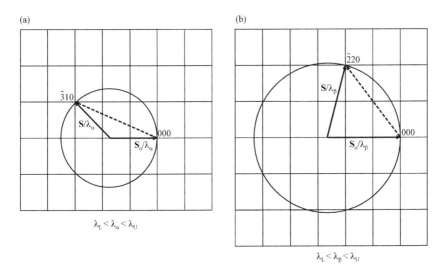

FIGURE 7.8 Examples of reflection in (a) the backward direction and (b) the forward direction.

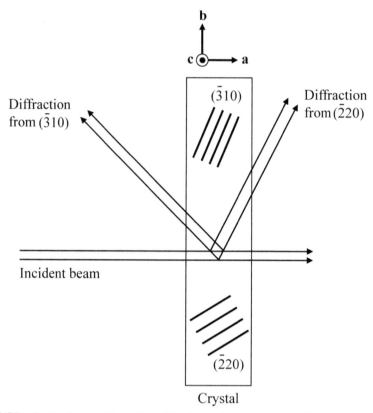

FIGURE 7.9 Real-space illustration of the forward and backward reflections.

The Laue method has two different configurations, depending on the relative locations of X-ray source, crystal, and film (Figure 7.10). In the *transmission Laue method*, the film is placed behind the crystal to record the beams diffracted in the forward direction. This method is so named because the recorded beams are transmitted through the crystal. In the *back-reflection Laue method*, the film is placed between the X-ray source and the crystal. Here, the incident beam passes through the center of the film and the beams diffracted in the backward direction are recorded by the film. In both cases, the diffracted beams form an array of spots on the film and the positions of these spots depend on the orientation of the crystal. Thus, either transmission or back-reflection method can be used to determine the orientation of a crystal. From the practical point of view, the back-reflection method is more dominantly utilized. Since the positions of diffraction spots on the film depend on the crystal orientation, it is necessary to orient the sample relative to the incident X-ray beam. Single crystals to be analyzed often have irregular shape and arbitrary size. In both configurations, the crystal is stationed on a sample holder (usually a goniometer) and can be rotated and tilted by adjusting it. In the transmission method, it is more or less inconvenient to handle the sample freely because the sample holder should not block the transmitted beams. In addition, when the sample has high absorption at X-ray wavelengths, it should be prepared very thin. Otherwise, the diffracted beams recorded on the film would be very weak. The back-reflection method, however, requires no special preparation of the sample, which may be of any shape and thickness. The back-reflection configuration is particularly useful for large crystals that absorb X-rays very strongly, for example, metal crystals.

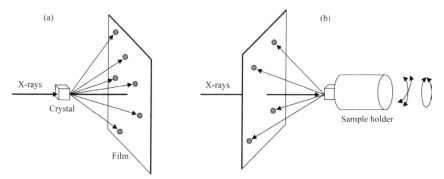

FIGURE 7.10 (a) Transmission Laue method. (b) Back-reflection Laue method.

The angle between the incident and diffracted beams is twice the Bragg angle. The back-reflection method records reflections of high Bragg angle, whereas the transmission Laue method captures only reflections of low Bragg angle. In other words, the back-reflected spots are obtained from the planes that make relatively large angles with the incident beam, while the transmitted spots are from those planes of small angles. Regarding the transmission Laue method, one obvious feature that is worthy of comment is the absence of reflections on the film over an area centered on the intersection of the incident beam with the film. A reflection close to the forward direction of the X-ray beam should have rather a small Bragg angle θ. Accordingly, $\lambda/2d$ for this reflection should also be small. As we have already seen in Figure 1.8, there is a sharp cut-off at the low wavelength side of the continuous emission from an X-ray tube, which depends on the operating voltage of the tube. This indicates that the wavelength required for this reflection may not be contained in the used white radiation. Moreover, the maximum value of d is limited by the unit cell dimensions of the crystal. Therefore, there will be a minimum value of θ below which reflection is impossible. It results in a circular blank area around the film center when the X-ray beam is incident parallel to one of the symmetry axes. Of course, the size and shape of the blank area on the film will depend on the orientation of the crystal relative to the incident beam. We can generally state that for a continuous spectrum with a certain cut-off wavelength, a substance with small unit cell dimensions (i.e., small d values) will exhibit a bigger blank area than a substance with larger cell dimensions. With more diffraction spots, it is easier to index the spots and determine the crystal orientation. The absence of reflections on the film is not a feature of the back-reflection Laue method. This is another reason why it is more widely used than the transmission method.

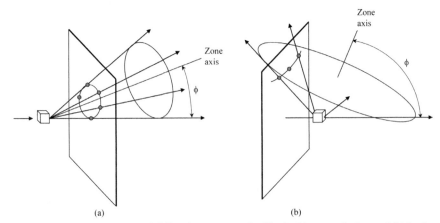

FIGURE 7.11 Formation of diffraction spots on the film in (a) transmission and (b) back-reflection geometry.

The diffraction spots obtained with either method can be seen to lie on certain curves. These curves are usually ellipses or hyperbolas in the transmission pattern and hyperbolas for the back-reflection pattern, as shown in Figure 7.11. The spots lying on a curve correspond to the reflections from planes belonging to one zone. It follows from the fact that the reflections from planes of a zone lie on the surface of a cone, whose semi-angle ϕ is the angle between the zone axis and the forward direction of the incident beam. Thus, one side of the cone is tangent to the transmitted beam. Diffractions along the cone sides will be represented by spots lying on a curve on the film. The shape of a zonal curve depends very much on the inclination angle φ of the zone axis to the transmitted beam. A film placed for the transmission geometry will intersect the cone in an ellipse, with the diffraction spots arranged on this ellipse (Figure 7.11(a)). If the angle φ does not exceed 45°, the cone will not intersect a film placed for the back-reflection configuration. When φ is between 45° and 90°, the cone intersects the film in a hyperbola, as shown in Figure 7.11(b). If φ equals 90°, the cone of diffracted beams becomes a plane containing the incident X-ray beam. In this case, the diffracted spots will be arranged on a straight line passing through the center of the film. Therefore, diffraction spots recorded on a back-reflection film lie on hyperbolas or straight lines. The distance of any hyperbola from the film center is a measure of the inclination angle of the corresponding zone axis.

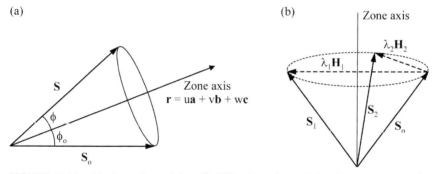

FIGURE 7.12 (a) Cone formed by all diffracting planes belonging to a zone. (b) Diffraction from two different set of planes that have the corresponding reciprocal lattice vectors $\mathbf{H}_1$ and $\mathbf{H}_2$.

The fact that the Laue diffraction from planes of a zone occurs in a conical manner can be easily demonstrated by considering the Bragg condition in reciprocal space. This is true regardless of whether the crystal has a symmetrical orientation or not. Consider a zone axis defined as $\mathbf{r} = u\mathbf{a} + v\mathbf{b} + w\mathbf{c}$, where $\mathbf{a}$, $\mathbf{b}$, and $\mathbf{c}$ are the unit cell axes of the crystal. Any planes belong to this zone if their indices (*hkl*) satisfy the relation,

$$hu + kv + lw = 0 \qquad (7.1)$$

because the plane normals are perpendicular to the zone axis; the dot product of $\mathbf{r} = u\mathbf{a} + v\mathbf{b} + w\mathbf{c}$ and $\mathbf{H}_{hkl} = h\mathbf{a}^* + k\mathbf{b}^* + l\mathbf{c}^*$ should be zero. From Eq. (4.9), the direction of diffraction from the (hkl) planes is given by

$$\mathbf{S} - \mathbf{S}_o = \lambda \mathbf{H}_{hkl} \qquad (7.2)$$

Forming the scalar product of $\mathbf{r} = u\mathbf{a} + v\mathbf{b} + w\mathbf{c}$ with Eq. (7.2), we have the following relation.

$$\mathbf{r} \cdot \mathbf{S} = \mathbf{r} \cdot \mathbf{S}_o + \lambda \mathbf{r} \cdot \mathbf{H}_{hkl} \qquad (7.3)$$

Since $\mathbf{r} \cdot \mathbf{H}_{hkl}$ is zero for all planes (hkl) belonging to this zone, we obtain $\mathbf{r} \cdot \mathbf{S} = \mathbf{r} \cdot \mathbf{S}_o$ and $\phi = \phi_o$, where ϕ_o is the angle between the incident beam and the zone axis, and ϕ is the angle between the zone axis and the diffracted beam from any (hkl) planes. As shown in Figure 7.12(a), the diffracted beams from all planes belonging to this zone form a cone that contains the incident beam as one element of the cone. Figure 7.12(b) is a graphical illustration of the diffraction from two different sets of planes that have the corresponding reciprocal lattice vectors $\mathbf{H}_1$ and $\mathbf{H}_2$. The unit vectors of the diffracted beams, given by $\mathbf{S}_1$ and $\mathbf{S}_2$, are equally inclined to the zone axis. While the diffracted beams have different wavelengths, their inclination angle to the zone axis is the same as the angle between the incident beam and the zone axis. If [uvw] is a prominent zone axis in the crystal, there are many reflections on the curve and reflections generated by planes of low indices will lie at the intersection of several such curves, each related to a zone axis. If the X-ray beam is incident along a symmetry axis of the crystal, the curves connecting the diffracted spots are also symmetric with respect to the center of the film.

As discussed in Section 5.4, the structure factors for hkl and $\overline{hkl}$ are of the same magnitude. It means that the reflections from either side of a set of (hkl) planes are equal in intensity: $I_{hkl} = I_{\overline{hkl}}$. When X-rays are incident parallel to a symmetry axis of the crystal, the resultant Laue pattern will exhibit the symmetry of that axis. However, it is important to note that the Laue pattern may not display all the symmetry elements possessed by the crystal, i.e., its point group symmetry. Suppose, for example, that the X-ray beam is incident parallel to the tetrad in a tetragonal crystal of point

Laue Method and Determination of Single Crystal Orientation 239

group 4/m and that the (hkl) plane is oriented so that it can reflect X-rays of wavelength λ. Then, the symmetry-related planes $\left(\overline{k}hl\right)$, $\left(\overline{hkl}\right)$, and $\left(k\overline{hl}\right)$ will also reflect X-rays of the same wavelength. All four reflections will have the same intensity and be symmetrically disposed on the film about its center. Similarly, when the incident beam is parallel to a hexad, triad, or diad axis in the crystal, the resulting Laue pattern will display the corresponding symmetry. If the incident beam is coplanar with a mirror plane in the crystal, the obtained pattern will exhibit a line of symmetry parallel to the mirror plane and passing through the center of the film. The Bragg law restricts θ to values between $0°$ and $90°$. Once the Bragg angle is set to one side of the (hkl) plane, there is no way of making a reflection occur from the other side of this plane. It is impossible to simultaneously record an hkl reflection and an $\overline{hkl}$ reflection without moving the crystal relative to the incident X-ray beam. In a tetragonal crystal of point group 4/m, (hkl) and ($\overline{hkl}$) are crystallographically identical due to the presence of a center of symmetry. However, if reflections are recorded from the planes (hkl), $\left(\overline{k}hl\right)$, $\left(\overline{hkl}\right)$, and $\left(k\overline{hl}\right)$, any reflections will not be observed from their opposite planes. Of course, if the crystal is rotated through $180°$ about an axis perpendicular to its tetrad and the opposite sense of the tetrad is brought into coincidence with the direction of the incident X-ray beam, reflections will take place from all four opposites. Then, the reflections from (hkl) and three symmetry-related planes will be absent.

The symmetry discernible on a Laue pattern is the crystal symmetry about a direction parallel to the incident beam. Since the mirror plane in a 4/m crystal is perpendicular to the tetrad, the mirror symmetry is not recorded on the film when the X-ray beam is incident parallel to the four-fold axis. This is the reason why the Laue pattern for a 4/m crystal will have the same symmetry as that for a tetragonal crystal of point group 4. Both have plane symmetry 4. By the same token, the diffraction pattern for a 4/mmm crystal will have plane symmetry 4mm, where the mirror planes parallel to the tetrad are recorded on the film as straight lines but the mirror plane perpendicular to it is not observed. The Laue symmetry of all tetragonal crystals is either 4 or 4mm. It is concluded that if the crystal is already known to be tetragonal, its plane symmetry can be determined by taking a single Laue photograph but the point group is not determined solely by it. This holds for other crystal systems. Figure 7.13 is a Laue pattern taken along the characteristics symmetry axis of $LiNbO_3$. As $LiNbO_3$ is a trigonal crystal with point group 3m, it shows a three-fold rotational axis and mirror planes vertically passing through the center of the film. Since LiNbO$_3$ lacks a center of symmetry, it does not possess a mirror plane perpendicular to the three-fold axis. Even if there were such a mirror plane, the observed diffraction pattern would have been the same as Figure 7.13, since the Laue method does not record any mirror plane perpendicular to the incident beam.

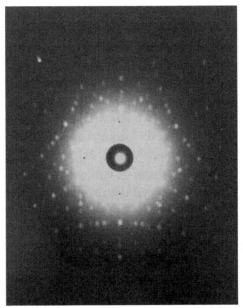

FIGURE 7.13   Laue diffraction pattern from a LiNbO$_3$ crystal.

7.3   INDEXING OF DIFFRACTION SPOT

The diffraction spots on a Laue photograph can be indexed, i.e., attributed to particular planes using special charts. Indexing can be achieved by correlating the angular relationships between the normals to reflecting planes with the known axial ratios and interaxial angles. The interpretation of back-reflection Laue patterns is facilitated by using the Greninger chart. The back-reflection Laue camera has a film-to-sample distance, L, of 3 cm or 6 cm. A Greninger chart for $L = 3$ cm is shown in Figure 7.14. This chart, usually printed on a transparency, is placed over the film to give the

Laue Method and Determination of Single Crystal Orientation   241

coordinates of the diffraction spots. The principle of this chart is explained with Figure 7.15, in which the X-ray beam is incident into the crystal situated at point O along the z-axis after passing through the film center denoted by point C. The x- and y-axes lie in the plane of the film and the beam diffracted by the plane shown is recorded at point S on the film. The normal to the diffraction plane intersects the film at point N. The incident beam, plane normal, and diffracted beam are coplanar and ON bisects the angle between the lines OS and OC. The diffraction plane is assumed to belong to a zone whose axis lies in the y-z plane. There may be some other planes that belong to this zone. The reflections from these planes will be recorded on the hyperbola AB, with their plane normals moving along the straight line DE. Thus, AB and DE are the trajectories of diffracted beams and plane normals on the film, respectively. To index any diffraction spot, it is necessary to know the orientation of the plane from which this spot comes. The angular coordinates γ and δ of the plane normal N can be easily correlated with the measured coordinates x and y of the diffraction spot S, once the film-to-sample distance is known. The result is the Greninger chart, which directly gives the angular coordinates γ and δ of the normal to the diffracting planes producing the spot. On the chart, the horizontal lines are curves of constant γ and the vertical lines are curves of constant δ.

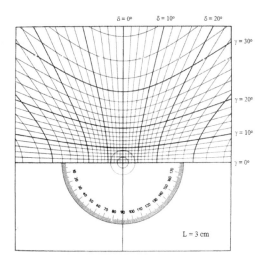

FIGURE 7.14   Greninger chart.

The Greninger chart shown in Figure 7.14 has 2° intervals both in the horizontal and vertical curves. The position of any diffraction spot S on this chart represents the γ and δ coordinates of the corresponding plane normal N. When reading the coordinates, the chart is placed over the film with its center consistent with the film center and with their edges parallel to one another. By rotating the film against the Greninger chart, we can get the angle between the reciprocal lattice vectors (i.e., the angle between the plane normals) of any two spots on the film. If we keep the center of the film on the Greninger chart center and rotate the film to bring the two spots on a curve of constant γ, then the difference in δ readings directly gives the angle between the two plane normals. The lower half of the chart has a protractor, which can be used to measure the rotation angle of the film. The Bragg angle of any diffraction spot, say, the spot S in Figure 7.15, can also be determined in a similar way. Rotating the film against the chart and bringing the point S on the curve of $\delta = 0°$, then the γ reading of this point equals the angle between the lines OC and ON in Figure 7.15. Let this angle be α. Since the Bragg angle θ is half the angle between the positive z direction and the line OS, we have $\theta = 90° - \alpha$. Nowadays, there are some other options except for the Greninger chart. One approach is to scan the Laue photograph and use a mathematical program or data analysis software to measure the angular coordinates of the spots. We can also use the software that generates a simulated Laue pattern.

There is no standard procedure for determining the Laue symmetry of a crystal and its orientation. From the practical point of view, the reflections on a Laue photograph cannot be easily indexed although the use of a Greninger chart or software is sure to be helpful. A common difficulty is that the Laue symmetry can be clearly displayed on the film only when the crystal is precisely aligned with its symmetry axis parallel to the incident beam. When two symmetry-related planes are inclined at different angles to the incident beam, the produced diffraction spots will be recorded at different distances from the center of the film. The two planes have the same interplanar spacing but will reflect different wavelengths, because the inclined angles to the incident beam (i.e., the Bragg angles) are unlike. The Laue method utilizes a wide range of continuous spectrum and the X-ray intensity may rapidly change in certain parts of the spectral range. Thus, the two spots may have significantly different strengths on the film even though the misalignment of the crystal is only marginal. When the crystals are of low symmetry, for example, monoclinic, indexing becomes

Laue Method and Determination of Single Crystal Orientation

very difficult due to the limited number of diffraction spots. Trial and error is inevitable until any conspicuous symmetry element is found. This is carried out by taking successive photographs while moving the sample-mounted goniometer relative to the direction of the primary X-ray beam. On the contrary, the Laue patterns from high-symmetry cubic and tetragonal crystals will contain a larger number of spots. Thus, one may be able to easily discern a trace of the symmetry axis at a considerable inclination to the incident beam (see Figure 7.3). Once the presence of a symmetry axis is conceived, the crystal should be adjusted to bring its orientation into precise alignment with the incident beam. Of course, a Laue photograph must be taken to confirm it.

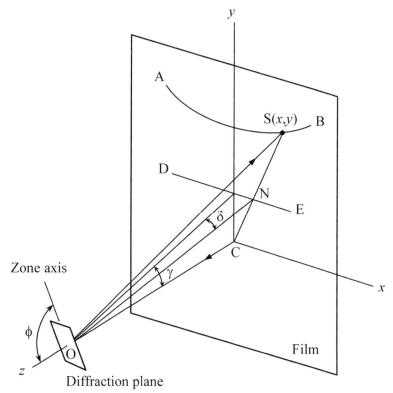

FIGURE 7.15 Positions of the diffraction spot and the normal to diffraction plane in back-reflection Laue method.

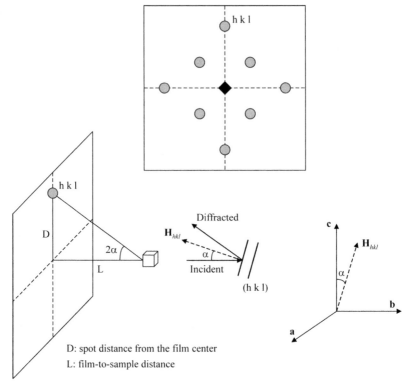

FIGURE 7.16 Indexing of an arbitrary diffraction spot in tetragonal system.

Except for specialized applications, it is rarely required to index all the diffraction spots. Some close to the film center suffice to determine and align the crystal orientation. Once the characteristic symmetry axis is determined, indexing can be carried out without using a Greninger chart. Assume that we have already aligned the **c**-axis of a tetragonal crystal parallel or anti-parallel to the incident X-ray beam through the procedure mentioned above. In either case, the tetrad is perpendicular to the film and passes through its center. The next step is to find how the **a**- and **b**-axes of the crystal are oriented with respect to the edges of the film. This is necessary when we wish to cut the crystal at right faces. Suppose that we like to index an arbitrary *hkl* spot on the imaginary Laue pattern shown in Figure 7.16, where the X-ray beam is incident anti-parallel to the **c**-axis of the crystal. The angle between the **c**-axis of the crystal and the diffracted beam, 2α, is given by

Laue Method and Determination of Single Crystal Orientation   245

$$\tan 2\alpha = D/L \qquad (7.4)$$

where D is the spot distance measured from the film center and L, the film-to-crystal distance. Thus, the angle α can be readily obtained. Since the reciprocal lattice vector $\mathbf{H}_{hkl}$ perpendicular to the diffracting planes bisects the angle between the **c**-axis and the diffracted beam, it makes an angle of α with the **c**-axis of the crystal. The dot product of $\mathbf{H}_{hkl}$ with the unit cell vector **c** gives the following relation

$$\cos\alpha = \frac{l/c}{\sqrt{h^2/a^2 + k^2/a^2 + l^2/c^2}} \qquad (7.5)$$

where a and c are the lattice parameters. The given spot can be indexed if we find out indices h, k, and l that satisfy this relation. The right Miller indices may be readily obtained by substituting some combinations of low indices into Eq. (7.5). In the tetragonal system, the **a**- and **b**-axes are interchangeable, i.e., indistinguishable due to the presence of a tetrad. It is obvious from the diffraction pattern that the **a**-axis of the crystal is either parallel to the edges of the film or 45° rotated from the edge. In the former case, the indices will be $h0l$ (or $0kl$), whereas the latter will have indices hhl.

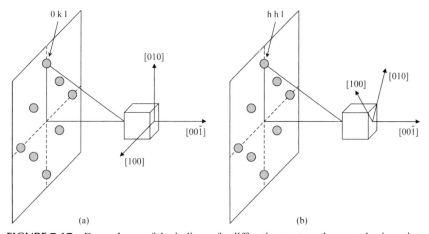

FIGURE 7.17 Dependence of the indices of a diffraction spot on the crystal orientation.

This is more clearly illustrated in Figure 7.17, where the X-ray beam is incident along the negative **c**-axis of a volume of tetragonal crystal; it is of cubic shape, not a cubic crystal. If the **a**-axis of the crystal, i.e., [100] is parallel to the horizontal edge of the film, the indices of the given spot will be 0*kl* (e.g., 013, 024, etc.) because the corresponding reflection should occur from planes running parallel to the **a**-axis. This results in $h = 0$. Conversely, if the indices of the spot were found to be 0*kl*, it means that the **a**-axis of this crystal was horizontally aligned (Figure 7.17(a)). Meanwhile, if the spot has indices *hhl*, the **a**-axis, i.e., [100] of the crystal would be 45° rotated from the horizontal edge, as shown in Figure 7.17(b). In this way, we can determine the in-plane orientation of the crystal projected onto the film. Figure 7.18 shows 013 and 113 reflections as specific examples. The diffraction spot to be indexed lies on a vertical line passing through the center of the film. To indicate this, the upper left corner of the film was cut in Figure 7.18. The incident beam, plane normal, and diffracted beam are always coplanar. The plane containing all of them intersects the film as a vertical line on which the diffraction spot is located. This plane is perpendicular to the **a**-axis of the crystal for the 013 reflection, as depicted in Figure 7.18(a). Thus, the **a**-axis will be parallel to the horizontal edge of the film. For the 113 reflection (Figure 7.18(b)), the plane containing the incident and diffracted beams is ($1\bar{1}0$), which will also cut the film as a vertical line. Therefore, the **a**- and **b**-axes of the crystal are at 45° from the edges of the film. Let's go back to the configuration of Figure 7.17(b), where the **a**- and **b**-axes of the crystal are at 45° from the edges of the film and its sides parallel to the incident beam are {110}. If we ultimately wish to prepare the crystal in {100} faces, it should be rotated through 45° about the film normal and then vertically cut.

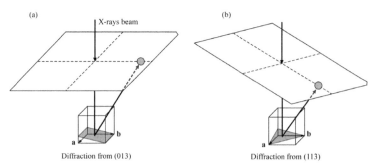

FIGURE 7.18 In-plane orientations of the crystal with respect to the film in (a) 013 reflection and (b) 113 reflection.

PROBLEMS

7.1. State the difference between the Laue diffraction and the electron diffraction. Is there a reflection(s), which is observable in one method but missing in the other?

7.2. Si has a lattice constant of 5.43 Å. A back-reflection Laue photograph is taken using a white X-ray beam impinging on a Si single crystal, where the film-to-sample distance is 3 cm.
  (a) Draw the expected diffraction pattern when the X-ray beam is incident along [001] of the crystal.
  (b) What is the closest distance of spots to the film center when the X-rays has a spectral range of 0.1–100 Å?
  (c) Describe how the results of (a) and (b) will change when the X-ray beam is incident along [111] of the same crystal.

7.3. Laue photography is very useful to determine the orientation of a crystal. However, it does not exhibit all the symmetry elements possessed by the crystal. Explain why?

CHAPTER 8

POWDER DIFFRACTION

CONTENTS

8.1   Introduction 250
8.2   Principle of Powder Diffraction 253
8.3   Indexing of Powder Pattern 256
8.4   Phase Identification 261
8.5   Determination of Crystal Structure 263
Problems 273

250 X-Ray Diffraction for Materials Research: From Fundamentals to Applications

8.1 INTRODUCTION

Powder diffraction is a technique using X-ray diffraction on powder or microcrystalline samples for the structural analysis of materials. It is most widely used for the phase identification of a crystalline material and can provide information on unit cell dimensions. As the name suggests, the sample is usually prepared in a powder form. The ideal sample consists of an enormous number of tiny crystal fragments in completely random orientation, as shown in Figure 8.1. Suitable samples may be obtained as fine-grained crystallites or by grinding crystalline materials. Here, "powder" means either an actual physical powder held together with a binder or any specimen in polycrystalline form. Thus, polycrystalline materials with a large number of small grains may be investigated nondestructively without the need for special sample preparation. This method involves the use of a monochromatic X-ray beam, which, in general, is the strong K_α characteristic line emitted from an X-ray tube. Copper is the most common target material, with Cu K_α radiation of λ = 1.542 Å. The K lines are classified with respect to the other energy level involved. While an electron transition from the L shell to the K shell gives rise to a K_α line, a transition from the M shell to the K shell produces a K_β line. Since the L shell is of lower energy than the M shell, the K_α line of a given element always has a longer wavelength than its K_β line. In general, the K_α line is much stronger than the K_β line because the vacant site of the K shell is more probably occupied by an L electron than by an M electron. Therefore, the K_α line is invariably selected when monochromatic X-radiation is required. For this purpose, the continuous emission from an X-ray tube is monochromated by a crystal monochromator or by a filter whose absorption edge falls between the K_α and K_β wavelengths. While the wavelengths of the characteristic lines depend only on the kind of the target, their intensities are influenced by the voltage applied across the tube. Below a certain threshold voltage, none of the accelerated electrons will have sufficient energy to eject a K electron from a target atom and no K lines will be emitted.

Powder Diffraction

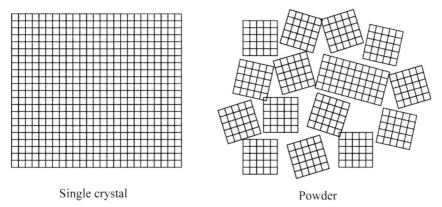

Single crystal Powder

FIGURE 8.1 A single crystal and its powder.

X-ray powder diffraction has been historically used for the identification and classification of minerals, but it can now be utilized for any materials as long as a suitable reference pattern is available. The powder diffraction allows for nondestructive analysis of multi-component mixtures without a special preparation procedure, which makes it possible to quickly analyze unknown as well as known materials encountered in such diverse fields as metallurgy, mineralogy, archeology, solid-state physics and chemistry, and biology. The primary use of powder diffraction lies in the identification and characterization of crystalline solids, each of which produces a distinctive diffraction pattern. Identification is performed by comparison of the obtained diffraction pattern to a known standard or to a database such as the Powder Diffraction File (PDF) of the International Center for Diffraction Data. In a powder pattern, both the peak positions and its relative intensities are characteristic of a particular phase and material. This provides a "fingerprint" for comparison. A multi-phase mixture will show multiple patterns superposed, enabling the relative concentrations of phases in the mixture to be determined.

The fundamental physics, upon which the method is based, provides high precision and accuracy in the measurement of interplanar spacings and unit cell dimensions, sometimes to fractions of an Ångström. The angular position of a diffraction peak is independent of the atomic arrangement within the unit cell and solely determined by the shape and size of the cell. Each peak represents a certain set of lattice planes and can therefore be characterized by Miller indices. If the crystal is highly symmetric, for example, cubic, it is not so difficult to index each peak even for an unknown phase. Once the peak is indexed, the unit cell dimensions may be derived from the obtained pattern. For the accurate determination of lattice parameters, a standard substance with precisely known cell dimensions can be added into the sample to correct its peak positions. The positions of diffraction peaks may be shifted by instrumental factors. It is not a rare case where the diffraction pattern measured today may be slightly different from that obtained from the same sample yesterday. Then, the standard substance is effectively utilized to calibrate the peak positions of the sample. A care is also required, particularly when the powder sample is prepared by grinding the crystal. A residual stress may be induced in the course of sample preparation and this is apt to alter the lattice parameters. The sample prepared in this way should be annealed at an appropriate temperature prior to diffraction experiment to remove any residual stress. The powder pattern can be recorded either on a photographic film or by a diffractometer. In the photographic method, the whole diffraction pattern is simultaneously recorded on a film, while it is scanned by a moving detector in the latter case. The resolution achievable in diffractometry is much better than in photography. The diffractometer is particularly useful when we need to determine the Bragg angles very accurately. It has an additional advantage that the positions and intensities of diffraction peaks can be measured simultaneously and quickly.

Although not impossible, the determination of an unknown crystal structure is extremely challenging due to the overlap of reflections. The crystal structure of a substance determines its diffraction pattern; the shape and size of the unit cell determines the angular positions of the diffraction peaks, and the atomic arrangement within the cell determines the relative intensities of the peaks. Thus, it should be possible to derive the structure from the pattern. However, it is not an easy task to directly deduce the structure from the observed pattern, since completely different structures may give rise to similar patterns. The general procedure is basically trial and error. A structure is first assumed to theoretically calculate its diffraction pattern and the calculated pattern is compared with the observed one. Complex structures require some mathematical programs as well as diffraction data. The determination of unknown crystal structures is very specialized, beyond the scope of this book. Here, the basic principle of structure determination is described with simple structures.

Powder Diffraction 253

8.2 PRINCIPLE OF POWDER DIFFRACTION

In X-ray diffraction, the incident beam, plane normal, and diffracted beam are always coplanar. Consider a reflection from (hkl) planes in a large single crystal, as shown in Figure 8.2. When a monochromatic X-ray beam is incident onto the crystal, it is diffracted only in a specific direction because the reciprocal lattice vector normal to the (hkl) planes has a fixed orientation and S_o, H_{hkl}, and S should be coplanar. In the powder method, the sample contains a very large number of tiny crystallites oriented at random with respect to the incident beam. Thus, each set of planes is randomly oriented. Some of the crystallites will be so oriented that their (hkl) planes satisfy the Bragg law and diffract the incident beam. The Bragg condition for powder diffraction may be better explained in reciprocal space. As the incident beam is monochromatic, the Ewald sphere has a fixed radius of $1/\lambda$. In the powder sample, every set of planes has random orientation, so does the set of (hkl) planes. Then, we can construct a sphere of reciprocal vector H_{hkl} whose radius is equal to the length of this vector (Figure 8.3). The incident beam vector, denoted by S_o/λ, terminates on the center of the H_{hkl} vector sphere. Two spheres intersect in a circle. A cone is formed by a number of lines connecting the center of the H_{hkl} vector sphere and the circumference of this circle. Reflection can occur from all (hkl) planes whose normals lie on this cone. The incident beam will thus be diffracted in a conical manner, giving rise to a cone of diffracted beams. The axis of the cone is coincident with the incident beam and makes an angle of 2θ with the diffracted beams. Figure 8.3 shows the diffraction condition for a particular set of planes. A crystal contains a number of sets of planes with different interplanar spacings. Different set of planes has different Bragg angles. The total diffraction pattern produced by a powder sample is thus a series of cones, each cone corresponding to a particular set of planes satisfying the Bragg law (Figure 8.4(a)). As we have described in Section 6.6, the grazing incidence X-ray diffraction is not suitable, if the film has a high degree of preferred orientation. On the contrary, a polycrystalline film has a number of small grains at random orientation. Thus, it diffracts the incident X-ray beam in a conical fashion, just like powder diffraction.

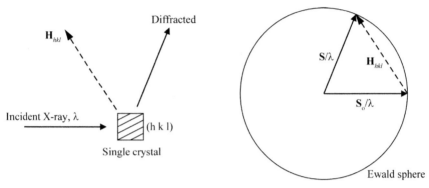

FIGURE 8.2 Diffraction from a single crystal.

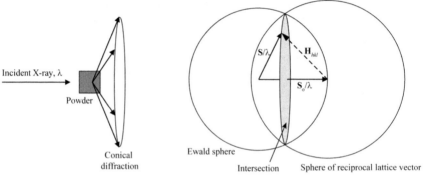

FIGURE 8.3 Powder diffraction in real and reciprocal space. Since a particular set of (*hkl*) planes is randomly oriented, diffraction occurs in a conical fashion.

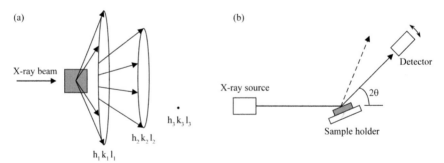

FIGURE 8.4 (a) Diffraction from different sets of planes. (b) Power diffractometer.

Powder Diffraction

The powder pattern has been recorded on a narrow strip of photographic film in a cylindrical camera whose axis is coincident with the specimen. The cones of diffracted beams intersect the cylindrical strip of film in lines and the resulting pattern appears when the film is unrolled and made flattened. Nowadays, the use of a diffractometer is more prevalent. The powder diffractometer consists of three basic elements: an X-ray tube, a sample holder, and a detector (Figure 8.4(b)). The powder sample is filled in a disc-shape container, with its surface carefully flattened. It is important that the sample should be homogeneous and large enough in area to catch the whole incident beam at the lowest Bragg angle to be employed. The container is put on one axis of the diffractometer and tilted by an angle θ with respect to the incident beam. The detector rotates around the sample holder on an arm at twice this angle. The intensity of diffracted X-rays is continuously recorded as the sample and detector rotate through their respective angles. Since the detector swings in a horizontal plane, it intercepts only a short arc in a cone of diffracted beams. In a powder sample, the plane orientations are completely random. Therefore, it is not necessarily required to vary the angle between the incident beam and the sample. However, it is more general to rotate the sample holder together with the detector. In the grazing incidence X-ray diffraction, the sample is stationary and only the detector is rotated. This is because if the sample is rotated, the path length of the X-ray beam through a thin film is drastically reduced.

Powder X-ray diffraction is mostly used to characterize and identify *phases*, and to refine already known structures rather than to solve unknown structures. In materials science and chemistry, it is often required to synthesize new materials. Although large single crystals are typically not immediately obtainable, micro-crystallites sufficient for powder diffraction may be readily available. Powder diffraction is therefore one of the most powerful methods to identify and characterize new materials in these fields. The unit cell parameters are somewhat temperature-dependent. Powder diffraction can be combined with *in situ* temperature control. As the temperature varies, the positions of the diffraction peaks will also change. This allows for the measurement of thermal expansion tensor of the material. When the material undergoes a phase transition, some diffraction peaks will newly appear or disappear. For instance, the diffraction peaks from (100) and (001) planes can be found at two different Bragg angles for a tetragonal crystal, while the two peaks will coincide in a cubic

256 X-Ray Diffraction for Materials Research: From Fundamentals to Applications

phase. Powder diffractometry can also be used in kinetic studies of polymorphic transitions. It the intensities of two diffraction peaks, one belonging to the reactant and the other to the product phase, are measured in a series of samples isothermally heated for different times, their ratio can be plotted against time to give the time required for a certain fraction of the reactant to be transformed into the product phase. Since the intensity of any diffraction peak is decreased by an increase in temperature, the exposure time required for a high-temperature diffraction experiment is rather long, compared to that used for typical room-temperature measurements. Atoms vibrate about their mean positions. This atomic vibration becomes more profound with increasing temperature. The diffraction peak is a consequence of the reinforcement of waves scattered at the Bragg angle. The Bragg law requires that the path length difference between waves scattered by atoms of adjacent layers be an integral multiple of the wavelength. As the temperature increases, the atoms vibrate more strongly. Therefore, the reinforcement is not as perfect as it is for a crystal with fixed atoms. This results in a decrease of the peak intensity. For a constant temperature, the diffraction intensity is more reduced by thermal vibration at higher Bragg angles (i.e., at smaller d-spacings) than at low angles.

8.3 INDEXING OF POWDER PATTERN

All substances produce a characteristic diffraction pattern. X-ray diffraction discloses the presence of a substance as its existing form, not in terms of the constituent elements. This is the reason why graphite can be differentiated from diamond by the diffraction method, though both consist of carbon. By the same token, X-ray diffraction reveals the presence of sodium chloride (NaCl), whereas typical chemical analysis detects only the presence of elements Na and Cl. Diffraction methods have the advantage that the substance does not have to be dissociated or dissolved. As stated in Chapter 5, the diffraction intensity is proportional to the squared magnitude of the structure factor. In powder diffraction, there is another factor strongly affecting the diffraction intensity, which is known as *the multiplicity factor*. All lattice planes of equal d-spacing give reflections at the same position, i.e., the same Bragg angle. Since reflections from such planes are independent of one another, the intensity of a powder peak will be simply the sum of the intensities of all reflections involved. A crystal may possess some symmetry elements and all lattice planes related by symmetry have the same d-spacing. Thus, the number of planes contributing to a powder reflection peak hkl will be the number of planes belonging to the form $\{hkl\}$. For example, all six planes of (100), (010), (001), ($\bar{1}$00), (0$\bar{1}$0) and (00$\bar{1}$) equally contribute to the 100 reflection of a cubic crystal. Thus, its multiplicity factor is 6. All these planes have different orientations in a single crystal but will be correctly oriented for powder diffraction with equal probability. Likewise, there are eight planes in the form $\{111\}$: (111), ($\bar{1}$11), (1$\bar{1}$1), ($\bar{1}\bar{1}$1) and their opposites. The corresponding intensity in a powder pattern will thus be eight times that of a single 111 reflection. (111) and its opposite ($\bar{1}\bar{1}\bar{1}$) planes may be crystallographically identical or different, depending on the point group of the crystal. Regardless of their surface properties, both planes are equally involved in diffraction and are counted separately in the multiplicity factor. The multiplicity factor also depends on the crystal system. In a tetragonal crystal, the (100) and (001) planes may have different d-spacings so that the multiplicity factor for $\{100\}$ is 4 and that for $\{001\}$ is 2. In brief, the multiplicity factor is the number of equivalent reflections that contribute to the powder peak.

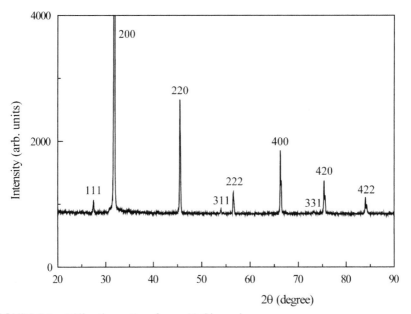

FIGURE 8.5 Diffraction pattern form a NaCl powder.

258   X-Ray Diffraction for Materials Research: From Fundamentals to Applications

The powder diffractometry enables a direct determination of the Bragg angle of every cone of diffracted X-rays. Solution of the Bragg equation for each diffraction peak gives the corresponding interplanar spacing, known as d-spacing. The spacing of the (hkl) planes in a cubic crystal is given by $d_{hkl} = a / \sqrt{h^2 + k^2 + l^2}$. Certain values of h, k, and l can be obtained for the values of $S = h^2 + k^2 + l^2$ calculated from the measured Bragg angles. Thus, if the unit cell dimension of the substance is known, the planes contributing to each diffraction peak can be easily indexed. In a cubic substance, some sets of planes that are not related by symmetry may have the same d-spacing so that their powder peaks are coincident; e.g., 300 and 221 peaks coincide with each other. Figure 8.5 shows a diffraction pattern of NaCl powder where Cu K_α radiation of $\lambda = 1.54$ Å was used as the monochromatic X-ray source. Each diffraction peak was indexed in Table 8.1. Since the lattice parameter of NaCl is already known ($a = 5.64$ Å), straightforward and unambiguous indexing can be achieved. Even if the unit cell is unknown, it is generally possible, with somewhat less certainty, to index the diffraction peaks and then to find the cell dimensions.

TABLE 8.1   Data From a NaCl Powder Pattern

Measured 2θ	$\sin^2\theta$	$S = h^2 + k^2 + l^2$	$h\,k\,l$	$\sin^2\theta/S = \lambda^2/4a^2$
27.46°	0.0563	3	111	1.877×10^{-2}
31.78°	0.0750	4	200	1.875×10^{-2}
45.53°	0.1497	8	220	1.871×10^{-2}
53.96°	0.2058	11	311	1.871×10^{-2}
56.62°	0.2249	12	222	1.874×10^{-2}
66.38°	0.3000	16	400	1.875×10^{-2}
73.20°	0.3555	19	331	1.871×10^{-2}
75.42°	0.3741	20	420	1.871×10^{-2}
84.10°	0.4486	24	422	1.869×10^{-2}

We here restrict our discussion mainly to cubic patterns that can satisfactorily be indexed at all times. For this purpose, suppose that we intend to index the NaCl pattern shown in Figure 8.5 without knowing its lattice parameter. Substitution of $d_{hkl} = a / \sqrt{h^2 + k^2 + l^2}$ into the Bragg equation yields

Powder Diffraction 259

$$\sin^2 \theta = \frac{\lambda^2}{4a^2}\left(h^2 + k^2 + l^2\right) = \frac{\lambda^2}{4a^2}S \qquad (8.1)$$

where S is an integer representing the sum of three squared Miller indices. This equation provides a means of indexing the powder pattern of any substance that is known to be cubic from other information or simply by suspicion. The values of $\sin^2 \theta$ can be calculated from the measured Bragg angles, which are also listed in Table 8.1. Since $\lambda^2/4a^2$ is a constant and S can have the values of 1, 2, 3, 4, 5, 6, 8, etc., the ratio between the $\sin^2 \theta$ values for two different peaks should be equal to the ratio between any two allowed integers. In Table 8.1, the ratio of the lowest $\sin^2 \theta$ value to the second-lowest value is very close to $0.75 = 3/4$. It means that the first peak (the smallest Bragg angle) has $S = 3$ and the second one, $S = 4$. That is, they have indices 111 and 200, respectively. Once the first two peaks are successfully indexed, the remainder can be indexed without difficulty, allowing the lattice parameter to be determined. Each peak was also indexed in Figure 8.5. Two things are conspicuous from the obtained powder pattern. The diffraction peaks with all even indices are much stronger than those with all odd indices. This is consistent with the structure factor of NaCl (Eq. (5.18)). As the Bragg angle increases, the overall peak intensity decreases. As we have already seen in Eq. (5.7), the intensity of the scattered radiation decreases with increasing scattering angle.

The derived $\lambda^2/4a^2$ value may slightly vary from peak to peak. The unit cell dimension can be more accurately evaluated when a measured value of θ as close as to $90°$ is used. Differentiation of the Bragg equation yields $2\sin\theta\Delta d + 2d\cos\theta\Delta\theta = 0$. Then

$$\frac{\Delta d}{\Delta\theta} = -d\cot\theta \qquad (8.2)$$

For a fixed error in θ, the deviation in d will be minimized as θ approaches $90°$. Measurement accuracy can be significantly improved by mixing the sample with a pure substance whose unit cell dimensions are very precisely known. This substance serves as an internal standard to calibrate the peak positions and then to reduce the error in θ. Si is the most common standard substance. Powder diffractometry is particularly useful for accurately measuring the unit cell dimensions of solid solutions. Cu and Au, both having an FCC structure, are completely mixed in the temperature range from $400°C$ to $900°C$ maintaining a single FCC phase.

260   X-Ray Diffraction for Materials Research: From Fundamentals to Applications

When the Cu-Au alloy is rapidly cooled, its lattice parameter varies linearly with atomic percentage from $a = 3.61$ Å for pure Cu to $a = 4.07$ Å for pure Au. Powder diffractometry on the quenched alloy will thus yield its composition.

Cubic crystals have only one unknown parameter, the unit cell edge a. The indexing of powder patterns becomes more difficult as the number of unknown parameters increases. In tetragonal crystals, there are two unknown parameters a and c. Then, the $\sin^2 \theta$ relation is given by

$$\sin^2 \theta = \frac{\lambda^2}{4a^2}\left(h^2 + k^2\right) + \frac{\lambda^2}{4c^2}l^2 = A\left(h^2 + k^2\right) + Bl^2 \qquad (8.3)$$

where A and B are constants. To solve this equation analytically, we need to find the value of A at first. For the $hk0$ peaks, Eq. (8.3) reduces to

$$\sin^2 \theta = A\left(h^2 + k^2\right) \qquad (8.4)$$

Since the allowed values of $h^2 + k^2$ are 1, 2, 4, 5, 8, ... etc, the $\sin^2\theta$ values of the $hk0$ peaks should be in the ratio of these integers. This procedure is to find out some peaks whose $\sin^2\theta$ values have the ratio of these permissible integers. Then the value of A can be obtained. B is obtained from the other peaks ($l \neq 0$) on the pattern. Equation(8.3) is rewritten as

$$\sin^2 \theta - A\left(h^2 + k^2\right) = Bl^2 \qquad (8.5)$$

The l^2 has the values of 1, 4, 9, 16, etc. Therefore, the left-hand side of Eq. (8.5) should have values in the ratio of these integers. The value of A is already known through the procedure described above. Various values of h and k are then assumed to find a consistent set of ratio; 1, 4, 9, etc. If these values are found, the value of B can be derived. Any orthorhombic crystals have three unknown parameters a, b, and c, and the indexing becomes far more difficult. The governing equation is in the form of $\sin^2 \theta = Ah^2 + Bk^2 + Cl^2$, where three unknown constants A, B, and C should be determined. Analytical methods of indexing the powder patterns are procedures to find certain numerical relationships among the experimentally observed $\sin^2 \theta$ values. The use of a computer may be inevitable to index the patterns of non-cubic crystals and many computer programs are currently available.

Powder Diffraction   261

8.4 PHASE IDENTIFICATION

The most widespread use of powder diffraction is in the phase identification of crystalline substances, each of which exhibits a distinctive diffraction pattern. Materials research often involves synthesis and modification of substances. When synthesizing or modifying a material, it is essential to confirm whether it has the desired phase. If an unknown phase (or pattern) is observed by chance, it is also necessary to identify what that is. A collection of diffraction patterns for a great number of different substances may allow identification of an unknown by recording its diffraction pattern and finding a pattern in the database file that exactly matches the pattern of the unknown. Hanawalt, a chemist who worked for Dow Chemical in the 1930s, was the first to realize the potential of creating a database. He and two colleagues began to collect and classify known diffraction patterns. Today, this activity is represented by the Powder Diffraction File (PDF) of the International Centre for Diffraction Data (ICCD). The 2006 PDF databases contained over 550,000 reference patterns. The PDF has many sub-files on elements, alloys, semiconductors, minerals, etc., with large collections of organic, inorganic, and organometallic compounds. The relevant activity has been carried out by the Joint Committee on Powder Diffraction Standards (JCPDS) found in 1969. The name of this organization was changed to the ICCD in 1978 to highlight its global commitment.

Most of the PDF data were obtained with Cu K_α radiation, except for those of Fe-containing substances. Fe is extremely absorbing at the wavelength of the Cu K_α line, making it difficult to retrieve much information from the substance. K_α consists, in part, of $K_{\alpha 1}$ and $K_{\alpha 2}$. $K_{\alpha 1}$ has a slightly shorter wavelength and is twice as strong as $K_{\alpha 2}$. These two lines are sufficiently close in wavelength. The powder patterns measured using the stronger $K_{\alpha 1}$ line only are also available for many substances. When they are not resolved as separate lines, a weighted average of the two lines, simply the Cu K_α line ($\lambda = 1.542$ Å), is used in calculation. The PDF databases are interfaced to a wide variety of diffraction analysis software and are searchable by computer. This information is typically an integral portion of the software that comes with the X-ray instrumentation. The card containing diffraction data is still called the JCPDS card among researchers. Note that a specific substance may have multiple cards because data have been progressively updated. Any diffraction pattern is characterized by a set of peak positions 2θ (also d-spacings) and a set of relative inten-

262   X-Ray Diffraction for Materials Research: From Fundamentals to Applications

sities where the maximum intensity is scaled to 100. If the strongest and weakest peaks in a powder pattern have absolute intensities of 200 and 16, their intensities are represented as 100 and 8 on the card, respectively. The reason why the d-spacing is listed on the card is because it is an invariant fundamental quantity, while the angular position of the peak depends on the used X-ray wavelength. As an example, a JCPDS card for $SrTiO_3$ is reproduced in Table 8.2. In the card, all observed peaks are arranged in sequence of increasing 2θ and decreasing d, being identified with the corresponding Miller indices.

TABLE 8.2 JCPDS Card For $SrTiO_3$ (PDF-2/Release 2001; International Centre for Diffraction Data) 35-0734

$SrTiO_3$	2θ (°)	d (Å)	Int.	hkl
Strontium Titanium Oxide	22.873	3.887	12	100
Tausonite, syn	32.424	2.760	100	110
Rad: $CuK_{\alpha 1}\lambda$: 1.5405 Å Filter: Ni Beta d-sp: Calculated	39.984	2.254	30	111
Ref: Swanson, H. Fuyat, Natl. Bur. Stand. (U.S), Circ. 539, 3, 44, (1954)	46.483	1.953	50	200
Sys. Cubic S.G. $Pm\bar{3}m$ (221)	52.357	1.747	3	210
a: 3.905 Å b: c:	57.794	1.595	40	211
α: β: γ:	67.803	1.382	25	220
	72.543	1.303	1	300
Pattern taken at 25°C. Sample from Nat. Lead Co.	77.175	1.235	15	310
Spectroscopic analysis: <0.01% Al, Ba, Ca, Si; <0.001% Cu, Mg.	81.721	1.178	5	311
Perovskite Super Group, 1C Group. PSC: cP5	86.204	1.128	8	222
Mwt: 183.52. Volume[CD]: 59.55	95.127	1.044	16	321

The PDF is a very powerful source for the identification of unknown substances. Identification of the unknown begins with obtaining its powder diffraction pattern. An ideal powder sample should have a great number of crystallites in random orientation. If the crystallites are very large,

Powder Diffraction 263

a smooth distribution of crystal orientations will not be achieved. They should be less than 10 µm in size to guarantee good powder statistics. Large crystallite sizes and non-random orientations both lead to peak intensity variation. Then, the obtained diffraction pattern may not agree with reference patterns in the PDF database. Computer-integrated diffractometers have software to catch the positions of peaks (also d-spacings) and to calculate their relative intensities. Phase identification can be performed by comparing a set of experimental d values with those in the database. A substance is characterized by the d-spacings of its three strongest peaks, referred to as d_1, d_2, and d_3. Since different substances may have nearly the same d-spacing for a specific peak, a set of d-spacings are necessary to characterize each substance. The three experimental values of d_1, d_2, and d_3, together with relative intensities, are usually sufficient to characterize an unknown pattern. Many 'search-match' programs can be used to compare experimental and tabulated values. Commonly this is an integral part of the software that comes with the instrumentation. Automated search/ match routine is first conducted for d_1, the d-spacing for the strongest peak, and is successively done for d_2 and d_3. When the closest match is found for d_1, d_2, and d_3, the d-spacings and relative intensities of all observed peaks should be compared with the tabulated values. If a complete agreement is achieved between the measured and reference patterns, the phase identification is finished. The identification of phases in a mixture is basically possible, but not an easy task unless the number of substances present in the mixture is two or three. The analysis becomes far more difficult when a diffraction peak from one phase overlaps a peak from another, and when this superimposed peak is one of the three strongest peaks in the unknown pattern. For reliable identification, it is essential to compare the powder pattern of the mixture with those of the suspected substances. The detection limit of any phase depends much on whether its diffraction pattern contains a very strong peak that can be resolved from the peaks of the other phases.

8.5   DETERMINATION OF CRYSTAL STRUCTURE

Now we briefly describe on the structure determination of a crystal from its diffraction pattern. The diffraction of X-rays by crystals was discovered by Laue in 1912. In the following year, Bragg revealed the structure of

NaCl, which was the first structural determination of crystalline material. Since then, the structures of innumerable crystals have been determined, including those containing a few hundreds of atoms within the unit cell. X-ray diffraction provides a primary means for the structural analysis of crystalline materials. Structures of increasing complexity have become soluble, since the techniques of structure determination have also progressed continuously. It is now possible to solve the crystal structures of biological proteins that contain thousands of atoms in the unit cell. Computer simulations based on the group theory and Fourier series approach are often utilized to reveal the possible atomic arrangement and are thus combined with X-ray and neutron diffractions in the determination of an unknown complex structure. These methods are out of the scope of this book. Although a complex structure cannot be determined solely by the X-ray diffraction, it is sure to be the most powerful and inevitable tool for the structural analysis. Here we are concerned only with the basic principle and its application to the solution of fairly simple structures of cubic symmetry.

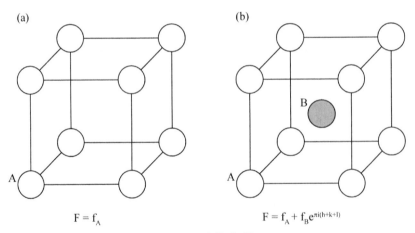

FIGURE 8.6 (a) Simple cubic structure and (b) CsCl structure.

The fundamental physics involved in structure analysis has already been discussed in Chapter 2, 4, and 5. In brevity, the shape and size of the unit cell determines the angular positions of the diffraction peaks, while their relative intensities are determined by the arrangement of atoms within the unit cell. In the determination of a crystal structure, the first step is to

Powder Diffraction

measure the unit cell dimensions. Once the shape and size of the unit cell is determined, the number of atoms per unit cell can be calculated from the measured density of the substance and its chemical composition. The atomic coordinates within the unit cell are finally deduced from the relative intensities of the diffraction peaks. As given in Eq. (8.1), the Bragg equation for a cubic crystal can be reformatted in the form of

$$\frac{\sin^2 \theta}{h^2 + k^2 + l^2} = \frac{\sin^2 \theta}{S} = \frac{\lambda^2}{4a^2} \tag{8.6}$$

where $S = h^2 + k^2 + l^2$ is an integer representing the sum of three squared Miller indices. Some integers, such as 7, 15, 23, 28, and 31, are impossible for the value of S because they cannot be produced as the sum of three squared integers. Since $\lambda^2 / 4a^2$ has a fixed value in the powder diffraction, $\sin^2 \theta / S$ is a constant for any one pattern while all the observed peaks have different $\sin^2 \theta$ values. Therefore, the indexing of a cubic diffraction pattern is to find a set of integers S that will yield a constant quotient satisfying this criterion. The possible set of integers S depends on the type of Bravais lattice; simple cubic, body-centered cubic (BCC), and face-centered cubic (FCC). Thus, we can figure out the Bravais lattice from the obtained set of integers S. If the observed diffraction peaks arise from a single, particular type of lattice, the $\sin^2 \theta / S$ value should be constant. It is to be noted that structure is different from lattice. Different structures may have the same type of Bravais lattice.

(a) 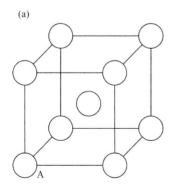 (b)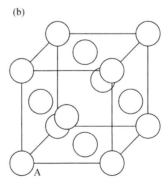

$F^2 = 4f_A^2$ if (h + k + l) is even $F^2 = 16f_A^2$ if h, k, and l are unmixed
$F^2 = 0$ if (h + k + l) is odd $F^2 = 0$ if h, k, and l are mixed

FIGURE 8.7 (a) BCC structure and (b) FCC structure.

266   X-Ray Diffraction for Materials Research: From Fundamentals to Applications

Figure 8.6(a) and (b) show simple cubic structure and CsCl structure, respectively. Both structures have a simple cubic lattice. In this primitive lattice, all planes have non-zero structure factor. That is, the simple cubic lattice has no systematic absence. Unlike the simple cubic structure, the CsCl structure has two different atoms per unit cell: one at (0,0,0) and the other at (1/2,1/2,1/2). Thus, if $h + k + l$ is odd, the structure factor will have a very small magnitude. Nevertheless, it is not zero because the two atoms are of different kinds. The set of integers S exhibited by a simple cubic lattice are listed in Table 8.3. Once the integers S are found, the hkl indices can be easily obtained by inspection. In some cases, the same value of S leads to more than one set of hkl indices.

TABLE 8.3   Set of Integers S Exhibited by a Simple Cubic Lattice

hkl	100	110	111	200	210	211	220	300	310	311	222	320
S	1	2	3	4	5	6	8	9	10	11	12	13

In the non-primitive lattices, such as BCC and FCC, the structure factors of certain lattice planes are zero and reflection from these planes is thus systematically absent. Figure 8.7(a) and (b) depict BCC and FCC structures, respectively. They are representative of the BCC and FCC lattices, since each atom forms one lattice point in both cases. The BCC structure has two atoms of the same kind per unit cell located at (0,0,0) and (1/2,1/2,1/2). As a result, the structure factor becomes zero when $h + k + l$ is odd. Therefore, such reflections as 100, 111, 210, and 300 are missing from the diffraction pattern. The corresponding set of integers S is then 2, 4, 6, 8, 10, etc., as listed in Table 8.4. This sequence of S values is common to all structures that have a BCC lattice. The structure factor of the FCC structure is zero when the h, k, and l indices are mixed. This leads to the set of S values of 3, 4, 8, 11, 12, 16, etc., as given in Table 8.5. Such other structures of FCC lattice as NaCl, GaAs, and α-ZnS also follow the same sequence. Diamond structure is an exceptional case. The diamond structure possessed by diamond, Si, and Ge also has an FCC lattice. Therefore, diffraction does not occur from the planes with mixed Miller indices. Even for unmixed indices, however, reflections are missing when the value of $h + k + l$ is an odd multiple of 2. This is because the diamond structure has two equivalent atoms associated with one lattice point. No diffraction peaks are observed from such planes as (200) and (222). In this structure, the S values have a sequence of 3, 8, 11, 16, etc.

TABLE 8.4 Set of Integers S in the BCC Structure

hkl	100	110	111	200	210	211	220	300	310	311	222	320
S		2		4		6	8		10		12	

TABLE 8.5 Set of Integers S in the FCC Structure

hkl	100	110	111	200	210	211	220	300	310	311	222	320
S			3	4			8			11	12	

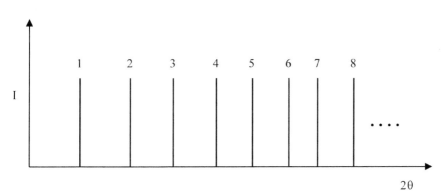

FIGURE 8.8 Hypothetical pattern.

The procedure of determining the Bravais lattice of a cubic substance and its lattice parameter can be better understood by introducing a hypothetical pattern and mentioning the steps required to index this pattern. Suppose that we have more than eight diffraction peaks from a cubic substance. The eight peaks, counted from the lowest 2θ angle, are numbered 1 to 8 in Figure 8.8. At this moment, we do not consider the peak intensities because the lattice type and dimension are dependent only on the angular positions of the peaks.

268 X-Ray Diffraction for Materials Research: From Fundamentals to Applications

TABLE 8.6 Hypothetical Diffraction Pattern from a Cubic Substance and Its Indexing

Peak #	2θ (°)	$\sin^2\theta$	$S = h^2 + k^2 + l^2$	$h\,k\,l$	$\sin^2\theta/S = \lambda^2/4a^2$
1	28.35	0.0600	1	100	6.000×10^{-2}
2	40.52	0.1199	2	110	5.995×10^{-2}
3	50.20	0.1799	3	111	5.997×10^{-2}
4	58.64	0.2398	4	200	5.995×10^{-2}
5	66.42	0.3000	5	210	6.000×10^{-2}
6	73.74	0.3600	6	211	6.000×10^{-2}
7	87.70	0.4799	8	220	5.999×10^{-2}
8	94.58	0.5399	9	300, 221	5.999×10^{-2}

TABLE 8.7 Hypothetical Diffraction Pattern that Leads to an FCC Lattice

Peak #	2θ (°)	$\sin^2\theta$	$S = h^2 + k^2 + l^2$	$h\,k\,l$	$\sin^2\theta/S = \lambda^2/4a^2$
1	27.46	0.0563	3	111	0.0188
2	31.72	0.0747	4	200	0.0187
3	45.00	0.1464	8	220	0.0183
4	53.90	0.2054	11	311	0.0187
5	56.54	0.2243	12	222	0.0187
6	66.33	0.2993	16	400	0.0187
7	72.98	0.3536	19	331	0.0186
8	75.25	0.3727	20	420	0.0186

Suppose that the peaks were observed at the 2θ values listed in the second column of Table 8.6. It is easy to find that their $\sin^2\theta$ values given in the third column are in the ratio of 1, 2, 3, 4, 5, 6, 8, and 9. The resulting set of integer S should be of the same sequence. This sequence is characteristic of a simple cubic lattice. The Miller indices for each peak can be deduced from the S value. Reflections 300 and 221 are superimposed for the peak with $S = 9$. If the observed peaks have the 2θ values listed in Table 8.7, the corresponding set of integers S becomes 3, 4, 8, 11, 12, 16, 19, and 20, revealing that the substance has an FCC lattice. Note that if the peaks corresponding to $S = 4$, 12, and 20 were absent from the list, it means that the substance has the diamond structure. This is a very special case where the crystal structure as well as the lattice type of a substance can be revealed only with the first indexing step. Whatever sequence the S values have, the lattice parameter can be derived from the relation of $\sin^2\theta/S =$

Powder Diffraction

$\lambda^2/4a^2$ and the used X-ray wavelength. According to Eq. (8.2), the error in "a" decreases as θ increases. Thus, the lattice parameter derived from the highest-angle peak is the most accurate. To calibrate any systematic error in θ, a standard substance whose unit cell dimensions are precisely known should be mixed with the sample in question. K_α radiation consists of a strong $K_{\alpha 1}$ line and a weaker $K_{\alpha 2}$ line. Although these two lines are sufficiently close in wavelength, the split of a diffraction peak into doublet is often observed under the K_α radiation, particularly at high Bragg angles. This makes it difficult to exactly define the peak position. While the whole K_α radiation is widely used for the purpose of phase identification, a filtered $K_{\alpha 1}$ line is highly preferred when the lattice parameter of a substance is to be accurately measured.

If the shape and size of the unit cell is successfully determined, we can calculate the volume of the unit cell. The next step is to find the number of atoms within the cell. This information should be known before the positions of atoms in the unit cell can be determined. The weight of all the atoms within the unit cell equals the density of the substance multiplied by the unit cell volume. The density ρ of a substance is given by

$$\rho = \frac{\Sigma W / N}{V} \tag{8.7}$$

where ΣW is the sum of the atomic weights of all atoms within the cell, N is Avogadro's number, and V is the volume of the unit cell. If the substance is an element of atomic weight M and contains n atoms in the unit cell, then

$$\Sigma W = N\rho V = nM \tag{8.8}$$

If the substance is a compound, then

$$\Sigma W = N\rho V = n'M_m \tag{8.9}$$

where n' is the number of "molecules" per cell and M_m, the molecular weight. For example, NaCl contains 4 Na atoms and 4 Cl atoms in the unit cell. Then, n' is 4 and $M_m = M_{Na} + M_{Cl}$, where M_{Na} and M_{Cl} are the atomic weight of Na and Cl, respectively. The constituent elements and composition of a substance can be easily revealed by ordinary chemical analysis. The volume of the unit cell can be calculated from its shape and size obtained in the previous step. Thus, the number of atoms in the unit cell can be determined if we measure the density of the substance.

The relative intensities of diffraction peaks are determined by the atomic arrangement in the unit cell. However, there is no general method of directly calculating the atomic positions from the observed intensities. We need to proceed by trial and error. As described previously, computer simulations based on the space group theory and Fourier series approach are often utilized in order to reveal the possible atomic arrangements. In case that the substance has a cubic lattice and the number of atoms per unit cell is not so large, we can approach it just by intuition. One example is given below. Let's suppose we are now to determine the atomic arrangements of an *AB* compound. We already know that this compound has an FCC lattice (through $\sin^2\theta$ values) and has 4 molecules in the unit cell (by the density measurement). How to determine the atomic arrangements of *A* and *B* atoms? There are two possible *AB*-type structures with FCC lattice: NaCl structure and ZnS structure. These two structures are depicted in Figures 8.9 and 8.10, respectively. Here, the structure of ZnS refers to that of its cubic phase, i.e., α-ZnS. Since the Bravais lattices of both structures are face-centered, their structure factor will be zero for the planes with mixed indices. This gives rise to an identical *S* sequence of 3, 4, 8, 11, 12, 16, 19, and 20.

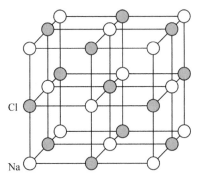

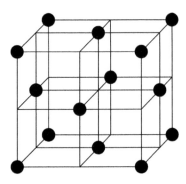

FIGURE 8.9 NaCl structure and its lattice.

Powder Diffraction

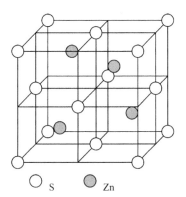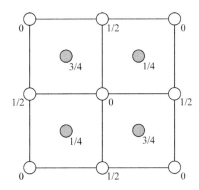

○ S ● Zn

FIGURE 8.10 ZnS structure and its (001)-projection.

We now turn to the relative intensities of the observed peaks. If the *AB* compound has the NaCl structure, its atomic coordinates within the unit cell are

A: (0, 0, 0), (½, ½, 0), (½, 0, ½), (0, ½, ½)

B: (½, 0, 0), (0, ½, 0), (0, 0, ½), (½, ½, ½)

Then, the structure factor will be

$$F^2 = 16 (f_A + f_B)^2 \quad \text{if } h, k, \text{ and } l \text{ are all even}$$

$$F^2 = 16 (f_A - f_B)^2 \quad \text{if } h, k, \text{ and } l \text{ are all odd} \quad (8.10)$$

If the *AB* compound has the ZnS structure, its atomic coordinates within the unit cell is then

A: (0, 0, 0), (½, ½, 0), (½, 0, ½), (0, ½, ½)

B: (¼, ¼, ¼), (¾, ¾, ¼), (¾, ¼, ¾), (¼, ¾, ¾)

The resulting structure factor will be

$$F^2 = 16 (f_A^2 + f_B^2) \quad \text{if } h, k, \text{ and } l \text{ are all odd}$$

$$F^2 = 16 (f_A + f_B)^2 \quad \text{if } (h + k + l) \text{ is an even multiple of 2}$$

$$F^2 = 16 (f_A - f_B)^2 \quad \text{if } (h + k + l) \text{ is an odd multiple of 2} \qquad (8.11)$$

Although the two structures have the same type of Bravais lattice, their intensity distributions will be totally different due to the unequal atomic configurations. For example, the 200 reflections will be strong in the NaCl structure while it is very weak and may not be detected for the ZnS structure. In Table 8.8, the intensity of each peak is qualitatively compared for both structures. There are many factors affecting the absolute intensity of a diffraction peak, including the output intensity of an X-ray tube, structure factor, multiplicity factor, polarization factor, etc. The relative intensities among the observed peaks, not their absolute intensities, are used to determine the atomic arrangement. The relative intensities of this AB compound will depend on whether it has the NaCl type structure or ZnS type structure. Figure 8.11(a) and (b) compare the intensity distributions expected when the compound has the NaCl structure and ZnS structure.

TABLE 8.8 Qualitative Comparison of the Diffraction Intensity of Each Peak in NaCl and ZnS Structures

Peak #	$S = h^2 + k^2 + l^2$	$h\,k\,l$	Intensity	
			NaCl structure	ZnS structure
1	3	111	Low	High
2	4	200	High	Low
3	8	220	High	High
4	11	311	Low	High
5	12	222	High	Low
6	16	400	High	High
7	19	331	Low	High
8	20	420	High	Low

Powder Diffraction

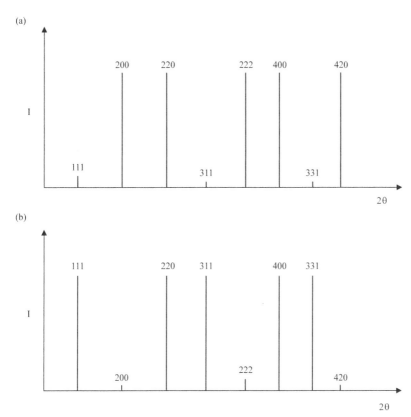

FIGURE 8.11 Expected patterns from (a) NaCl structure and (b) ZnS structure.

PROBLEMS

8.1. A material is known to be cubic, with one atom per lattice point. Through the powder diffraction experiment, the five largest interplanar spacings are found to be: 2.087 Å, 1.808 Å, 1.278 Å, 1.090 Å, and 1.044 Å. Determine the Bravais lattice of this material and its lattice parameter.

8.2. A powder diffraction pattern was obtained from a crystal of BCC lattice (a = 3.154 Å), which was conducted in the 2θ range of 20°–120° using a Cu K$_\alpha$ line at 1.542 Å. Specify the angular positions and indices of the observed peaks.

274   X-Ray Diffraction for Materials Research: From Fundamentals to Applications

8.3. Powders of two different elements **A** and **B** are mixed together and a powder diffraction experiment was performed with this mixture. **A** has an FCC structure with a = 4 Å and **B** is body-centered tetragonal (a = 4 Å and c = 6 Å). Cu K_α line is used as the X-ray source and 2θ is in the range of 20°–120°. State the positions and indices of expected peaks.

8.4. Powder diffraction experiment using a Cu K_α line was carried out for a material in the range of 2θ = 20°–100°. The stable structure of this material is BCC with a = 3.42 Å.

(a) State the positions of the observed peaks and from which planes they are coming.

(b) Let's assume that the material is transformed into FCC structure by application of a high pressure. Then, repeat "(a)".

8.5. Compare the powder diffraction patterns expected from Si and GaAs.

APPENDIX: FOURTEEN BRAVAIS LATTICES

There are five different types of two-dimensional (2D) lattices, simply called plane lattices: *oblique P-lattice, rectangular P-lattice, rectangular C-lattice, square P-lattice,* and *hexagonal P-lattice.* In crystallographic notations, P, C, I, F represent "primitive", "base-centered", "body-centered", and "face-centered", respectively. Figure A1 shows five 2D lattices along with their rotational symmetry elements. The most general 2D lattice (a ≠ b, γ general) has a diad perpendicular to the plane through every lattice point and midway between lattice points. This lattice type is known as oblique P-lattice, since the unit cell contains one lattice point. The rectangular P- and C-lattices also have diads only, but their unit cells are rectangular with $\gamma = 90°$. When the two unit cell axes of the rectangular P-lattice are of equal length, the square P-lattice is obtained. This type of lattice possesses four-fold rotational symmetry. The hexagonal P-lattice has a hexad through every lattice point, a triad on the center of a triangle formed by three lattice points, and a diad midway between two adjacent lattice points.

The 'Fourteen Bravais Lattices' can be derived from the five 2D lattices. To build up a space lattice, we should stack these 2D nets regularly above one another to form an infinite set of parallel sheets. We start with the net based on a parallelogram, i.e., the oblique P-lattice of Figure A1. If we stack nets of this type so that the lattice points in successive nets do not lie vertically above one another (Figure A2), the two-fold rotational symmetry is not maintained. Then, we have a *simple triclinic lattice* that exhibits no rotational symmetry. The unit cell is an arbitrary parallelepiped with edges *a, b, c,* no two of which are necessarily equal. The unit cell angles can take any value. In order to preserve two-fold symmetry, the two-fold axes in successive nets should be coincident with one another. There are two different ways of accomplishing this. We can arrange nets vertically above one another, or we can produce the staggered arrangement, viewed perpendicular to the nets, as shown in Figure A3. In the staggered arrangement, the two-fold axes at the unit cell corners of the second net coincide with those at the centers of the unit cell sides of the first net. These two stacking sequences produce the simple monoclinic and base-

276   X-Ray Diffraction for Materials Research: From Fundamentals to Applications

centered monoclinic lattices, both of which exhibit the two-fold rotation symmetry along one direction.

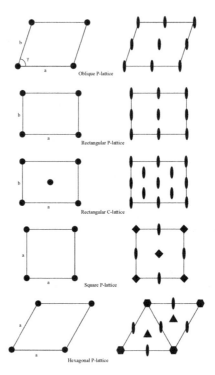

FIGURE A1   Five 2D lattices and their rotational symmetry elements.

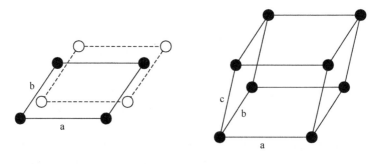

Simple triclinic (P)

FIGURE A2   Buildup of a simple triclinic lattice.

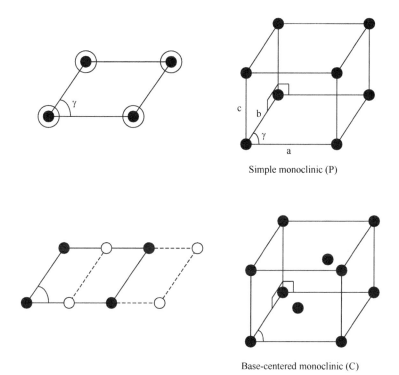

FIGURE A3 Stacking sequences for monoclinic P and C lattices.

Orthorhombic system, which is characterized by three mutually perpendicular diads, can have four different types of space lattices: simple, body-centered, based-centered, and face-centered. Two of them are derived from the rectangular P-lattice, and the other two, from the rectangular C-lattice. If we stack rectangular P-nets vertically above one another, we produce the simple orthorhombic lattice, as shown in Figure A4(a). The unit cell, a rectangular parallelepiped, possesses three mutually orthogonal diads. We can also preserve the orthorhombic symmetry by stacking the rectangular P-nets in a staggered sequence, as shown in Figure A4(b). The lattice obtained by this sequence is the body-centered orthorhombic. If we stack rectangular C-nets vertically above one another, we produce the base-centered orthorhombic lattice (Figure A4(c)). Another staggered arrangement of the rectangular C-nets leads to the face-centered orthorhombic lattice, as depicted in Figure A4(d).

278   X-Ray Diffraction for Materials Research: From Fundamentals to Applications

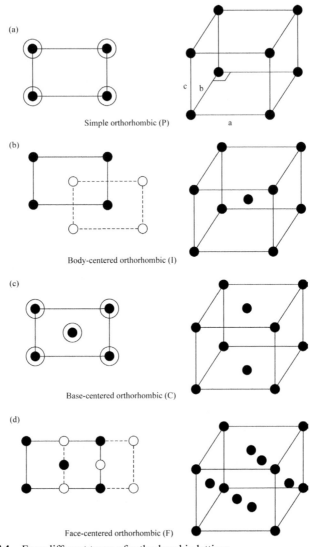

FIGURE A4 Four different types of orthorhombic lattices.

The two tetragonal lattices can be easily developed. The square net in Figure A1 has four-fold rotation axes at the corners and centers of the squares. The four-fold symmetry can be preserved if the second net is stacked over the first net, with their corners vertically overlapping one another. This four-fold symmetry is also maintained by placing the square corner of the second net above the square center of the first net. The resulting space lattices are the simple tetragonal and body-centered tetragonal lattices shown in Figure A5.

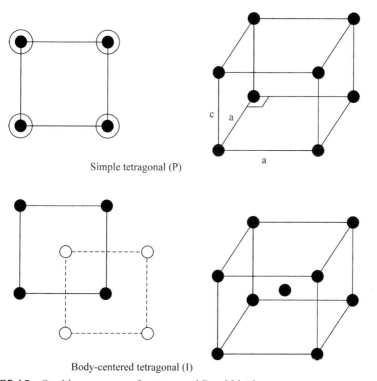

FIGURE A5 Stacking sequences for tetragonal P and I lattices.

We have so far derived nine of the Bravais lattices. All the remaining lattices are based on the stacking of the hexagonal P-lattice. To preserve six-fold rotational symmetry in a space lattice, such hexagonal nets should be stacked vertically above one another, as shown in Figure A6. There is no other way of maintaining the six-fold symmetry. Therefore, hexagonal system has a single lattice type: simple hexagonal. A space lattice consistent with three-fold rotational symmetry can be obtained by stacking these hexagonal P-nets in a staggered fashion. As shown in Figure A1, the three-fold rotation axes run through the centers of triangles formed by the lattice points. Thus, if the second net is stacked in such a way that its lattice points are placed above the centers of either upright or inverted triangles of the first net, the three-fold symmetry is maintained, as illustrated in Figure 2.13(a). To meet the fundamental translational symmetry (i.e., periodicity) of a lattice, the third net should be stacked by the same fashion. If stacking proceeds in this way, the fourth net overlaps the first one when viewed vertically. The primitive cell of the constructed lattice is a rhombohedron as depicted in Figure 2.13(b). The trigonal system also has a single lattice type, although a triple hexagonal cell given in Figure 2.13(c) is often taken as the conventional unit cell. As we have already described in Chapter 2, three different types of cubic lattices can be built up from the same 2D lattice by adjusting the height between the successive nets.

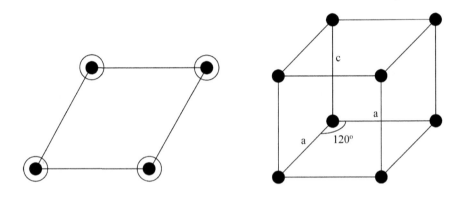

Simple hexagonal (P)

FIGURE A6 Hexagonal system has a single lattice type: simple hexagonal.

BIBLIOGRAPHY

1. Hecht, E. *Optics, 4th ed.*; Addison-Wesley, 2001.
2. Pedrotti, F., Pedrotti, L. M., Pedrotti, L. S. *Introduction to Optics, 3rd ed.*; Addison-Wesley, 2006.
3. Griffiths, D. *Introduction to Electrodynamics, 4th ed.*; Addison-Wesley, 2012.
4. Cheng, D. Field and Wave electromagnetics, 2nd ed.; Addison-Wesley, 1989.
5. Saleh, B. E., Teich, M. C. *Fundamentals of Phtonics, 2nd ed.*; Wiley, 2007
6. *International Tables for X-ray Crystallography, vol. 3, 2nd ed.*; 1968.
7. Kelly, A., Knowles, K. *Crystallography and Crystal Defects, 2nd ed.*; Wiley, 2012.
8. McKie, D., McKie, C. *Essentials of Crystallography*; Blackwell Scientific Publications, 1986.
9. Gaponenko, S. *Introduction to Nanophotonics*; Cambridge, 2010.
10. Novotny, L., Hecht, B. *Principles of Nano-Optics*;Cambridge, 2006.
11. Fowles, G. *Introduction to Modern Optics, 2nd ed.*; Dover, 1989.
12. Cullity, B. D., Stock, S. R. *Elements of X-ray Diffraction, 3rd ed.*; Prentice Hall, 2001.
13. Warren, B. E. *X-Ray Diffraction*; Dover, 1990.
14. Suryanarayana, C., Grant Norton, M.*X-RAY Diffraction: A Practical Approach*; Kluwer Academic/Plenum Publishers, 1998.
15. Jenkins, R., Snyder, R. L. *Introduction to X-ray Powder Diffractometry*; John Wiley & Sons, Inc, 1996.
16. Hammond, C. *TheBasics of Crystallography and Diffraction*; Oxford University Press, 1997.
17. Goodhew, P., Humphreys, J., Beanland, R. *Electron Microscopy and Analysis, 3rd ed.*; Taylor & Francis, 2001.
18. Williams, D., Carter, C. *Transmission Electron Microscopy, 2nd ed.*; Springer, 2009.
19. Bozzola, J., Russell, L. *Electron Microscopy, 2nd ed.*; Jones & Bartlett Publishers, 1998.
20. Yariv, A., Yeh, P. *Optical Waves in Crystals*; Wiley-Interscience, 2002.
21. Kittel, C. *Introduction to Solid State Physics, 8th ed.*; Wiley, 2004.
22. Degiorgio, V., Cristiani, I. *Photonics: A Short Course (Undergraduate Lecture Notes in Physics)*; Springer, 2014.
23. Pearsall, T. *Photonics Essentials, 2nd ed.*; McGraw-Hill Professional, 2009.
24. Rhodes, G. *Crystallography Made Crystal Clear, 3rd ed.*; Academic Press, 2006.
25. Jens Als-Nielsen, Des McMorrow. *Elements of Modern X-ray Physics, 2nd ed.*; Wiley, 2011.
26. Ewald, P. P. Introduction to the dynamical theory of X-ray diffraction. *Acta Crystallographica Section,* 1969,A25, 103.
27. Bragg, W.H., Bragg, W.L. The Reflexion of X-rays by Crystals. *Proc R. Soc. Lond.*1913, A88 (605), 428–38.

28. Jach, T., Cowan, P.L., Shen, Q., Bedzyk, M.J. Dynamical diffraction of X-rays at grazing angle. *Phy. Rev. B*, 1989, 39, 5739.
29. Nauer, M., Ernst, K.,Kautek, W., Neumann-Spallart, M. Depth Profile Characterization of Electrodeposited Multi- Thin-Film Structures by Low Angle of Incidence X-Ray Diffractometry. *Thin Solid Films*, 2005, 489, 86-93.
30. Bube, R. *Electron in Solids, 3rd ed.*; Academic Press, 1992.
31. Campbell, M., Sharp, D., Harrison, M., Denning, R., Turberfield, A. Fabrication of photonic crystals for the visible spectrum by holographic lithography. *Nature*, 2000, 404, 53.
32. Yang, B., Lee, M. Laser interference-driven fabrication of regular inverted-pyramid texture on mono-crystalline Si. *Microelectronic. Eng.*, 2014, 130, 52.
33. Yang, B., Lee, M. Mask-free fabrication of inverted-pyramid texture on single-crystalline Si wafer. *Optics &Laser Technol.*, 2014, 63, 120.
34. Shin, H., Yoo, H., Lee, M. Fabrication of Au thin film gratings by pulsed laser interference. *Appl. Surf. Sci.*, 2010, 256, 2944.
35. Hsieh, M., Lin, S. Compact holographic lithography system for photonic-crystal structure. *J. Vac. Sci. Technol. B*, 2011, 29, 011015.

INDEX

A

Absorption, 10, 13, 15, 19–25, 156, 235, 250
Absorption edge, 19, 24, 156
Analytical methods of,
  indexing powder patterns, 260
Angle of incidence, 95, 104, 107, 120, 121, 217, 218
Ångström fractions, 251
Atomic arrangements, 43
  (110) plane, 71
  atomic planes, 74
Atomic scattering factor, 155–157, 162, 176
Avogadro's number, 269

B

Back-reflection method, 234, 235
Backward direction reflection, 232–234
Base-centered lattice, 30
Beryllium, 15
Binary alloy system, 83
Body-centered cubic (BCC)lattice, 30, 42, 66, 52, 75, 77, 149, 150, 173, 174, 189, 265, 266, 273
Bragg law, 51, 102, 119–131, 145, 148–150, 157, 158, 162, 170, 172, 190, 195, 201, 207, 226, 230, 233, 239, 253, 256
  Bragg angle, 51, 121, 131, 134, 136, 148, 171, 214, 217, 226, 231, 235, 239, 242, 255, 256, 258, 259
  Bragg equation, 121, 128, 226, 258, 259, 265
  Bragg peak, 119, 156, 214
  Bragg reflection, 121
Bravais lattices, 38, 39, 66, 270, 279

C

Center of symmetry, 34, 48, 50, 88, 91, 172, 239
Characteristic lines, 17–19, 250
Characteristic radiation, 11, 17, 19, 24
Coherent interface, 208
Coherent scattering, 24, 26, 153
Compton scattering, 23, 154
Constructive interference, 99, 100, 104, 111, 127, 134, 182
Continuous radiation, 16, 17
Continuous spectrum, 10, 19, 226, 229, 236, 242
Crystal geometry, 59
Crystal orientation, 187, 229, 230, 235, 236, 244, 245
Crystal structure, 4, 5, 28, 29, 31, 37, 38, 52, 66, 67, 72, 86, 119, 145, 148, 149, 173, 182, 252, 264, 268
  body-centered cubic, 66
  face-centered cubic, 66
  hexagonal close-packed, 66
Crystal system, 28, 31, 35–38, 42–44, 49, 58, 61, 73, 86, 90, 148, 239, 257
  cubic, 35
  hexagonal, 35
  monoclinic, 35
  orthorhombic, 35
  tetragonal, 35
  triclinic, 35
  trigonal, 35
Crystalline, 4, 24, 26, 28, 31, 64, 78, 109, 119, 138, 139, 143, 177, 182, 184, 185, 191, 196, 200, 201, 214, 215, 218, 222, 250, 251, 261, 264
Crystallographic, 45, 74, 88, 89, 199, 239, 257
  notations (P, C, I, F), 275

284   X-Ray Diffraction for Materials Research: From Fundamentals to Applications

base-centered, 275
body-centered, 275
face-centered, 275
primitive, 275
orientations, 185
planes, 46
Crystallography, 28, 29, 86, 128
Crystals, 32, 35, 37
CsCl structure, 81, 82, 149, 150, 164, 165, 264, 266
Cubic system, 30, 35–39, 42, 44, 45, 52, 56, 57, 60–62, 67, 75, 81–86, 88–90, 125, 133, 140, 145, 149, 150, 167, 168, 172–176, 189–193, 196, 199, 209, 221, 222, 226, 243, 246, 252, 255–270, 273, 280
Cubic crystal, 38, 192

D

De Broglie's hypothesis, 141
    postulate, 141
Definition of, refractive index, 94
Destructive interference, 98–100, 104, 118, 134, 138, 149, 157
Detector, 124, 133, 187–189, 195, 196, 214, 217–219, 252, 255
Determination of,
    crystal structure, 263
    in-plane orientations, 191, 193
Diamond structure, 78, 80, 90, 91, 166, 167, 174–176, 212, 266, 268
Diffraction, 4–8, 13, 18, 24, 26, 28, 51, 52, 84, 105–110, 118–163, 166–177, 182, 183, 187, 189–191, 195–200, 208, 210, 214–223, 226–274
    geometry, 133
    Huygens's principle, 105
    intensity, 52, 120, 134, 145, 149, 157, 166, 169, 189, 216, 222, 256
    pattern, 190, 199, 257
    spots, 118, 142, 172, 174, 229, 237, 240
Diffractometer, 124, 186–188, 252, 254, 255
Disc-shape container, 255

E

Elastic deformation, 203, 204
Electric field, 6, 7, 10–12, 24, 50, 51, 109, 110, 151–154
Electromagnetic radiation, 4, 5, 8, 11–13, 20, 21, 96, 118
Electromagnetic spectrum, 5, 9, 11, 96
    sinusoidal function, 5
    wave with oscillating electric and magnetic fields, 5
Electromagnetic theory, 9, 151
Electromagnetic waves, 4–10, 12, 13, 20, 24, 95–99, 108, 114, 118, 151, 152
    frequency, 7
        temporal period, 7
    wavelength, 7
        spatial period, 7
Electron diffraction, 51, 141–144, 174, 247
Electro-optic effect, 50
Epitaxial film, 212–215, 217–219
Ewald construction, 130, 131
Ewald sphere, 130–132, 141, 142, 174, 231, 232, 253

F

Face-centered cubic (FCC) lattices, 30, 42, 52, 61, 64, 66–68, 76, 78, 82, 84, 165–167, 172, 175, 265, 266, 268, 270
    structure, 67
Face-centering translations, 165, 166, 173
Fermat's principle, 95, 114
Film quality, 210
Film/substrate structure, 113
Fingerprint, 182, 251
First-order reflection, 122, 123
Fluorescent radiation, 24
Forward direction, 154, 155, 156, 233–235, 237
    reflection, 233
Four-circle diffractometer, 186
Fourier series approach, 7, 264, 270
Fourteen Bravais lattices, 275
    five 2D lattices, 275

Index

Full width at half maximum (FWHM), 137
Fundamental translational symmetry, 40, 280
   periodicity of crystals, 40

G

Glass prism, 94
Gnomonic projection, 227, 228
Goniometer, 186, 235, 243
Grating period, 4, 108
Grazing incidence in-plane X-ray diffraction (GIIXRD), 219, 220
Grazing incidence x-ray diffractions, 216
Greninger chart, 240–242, 244
Group theory, 28, 264, 270
Γ-rays, 8, 11

H

Hexagonal close-packedstructures, 66, 72, 167
Hexagonal P-lattice, 275
Hexagonal system, 35–41, 49, 50, 56, 57, 60, 61, 64, 66, 72, 73, 79–81, 85, 90, 91, 167, 177, 220, 276, 279, 280
Hooke's law, 204
Huygens's principle, 105, 106
Hyperbolas, 236, 237
Hypothetical diffraction pattern, 268
   cubic substance and its indexing, 268
   leads to FCC lattice, 268
Hypothetical pattern, 267
Hypothetical points, 29, 31

I

Incoherent interface, 208
In-plane orientation, 183, 184, 191, 192, 195, 196, 198–200, 218, 222, 246
Incoherent radiation, 23, 154
Incoherent scattering, 23
Indexing, 47, 49, 142, 242, 244, 258–260, 265, 268
   powder pattern, 256
Indices of, 45, 46

direction, 45
   directions of a form, 45
plane, 46
   crystallographic planes, 46
   planes of a form, 48
Induced electrostatic force, 11
Infrared, 8, 25, 114
Integrated intensity, 171
Interference, 93, 96, 100, 101, 104, 109, 126
   acoustic waves, 96
   effects of interference, 96
   electromagnetic waves, 96
   fringe pattern, 100
   matter waves, 96
   surface waves, 96
Internal stress, 4, 5, 183
International Centre for Diffraction Data, 261, 262
Intrinsic stresses, 208
   growth stresses, 208
Inverse spinel structures, 82
Inverted tetrahedron, 71

J

Joint Committee on Powder Diffraction Standards, 261

K

Kinetic energy, 15, 16, 23, 25, 154

L

Lattice parameters, 29, 38, 41, 44, 45, 60, 63, 89, 144, 194, 195, 207, 208–210, 221, 245, 252
Lattice plane, 58, 79, 172
Lattice point, 29–31, 43, 44, 47, 56, 66, 69, 72, 75, 78–82, 130, 139, 151, 165, 166, 167, 171, 174, 215, 232, 266, 273, 275
Lattice types, 30, 31, 38, 44
   base-centered, 30
   body-centered, 30
   face-centered, 30
   simple, 30

286 X-Ray Diffraction for Materials Research: From Fundamentals to Applications

Laue colored pattern, 230
Laue diffraction, 227, 228
Laue equations, 127, 128
Laue method, 18, 226, 229, 230, 233–236, 240, 242, 243
Laue photograph, 239, 240, 242, 243, 247
Linear absorption coefficient, 20, 22
Lorenz force microscopy, 144

M

Mass absorption coefficient, 22–24
Mass thickness, 22
Maxwell equations, 9, 11
Microcrystalline samples, 250
Microwave oven, 10
Microwaves, 9, 10
Miller indices, 46–50, 56, 58, 90, 128, 142, 162, 164, 166, 174, 183, 185, 245, 251, 259, 262, 265, 266, 268
 plane designation, 46
Miller-Bravais indices, 49, 50
Misfit dislocation, 210, 212
Molybdenum target, 16, 17
Monochromator, 20, 139
Monoclinic, 35, 36, 39, 89, 242, 275–277
Mosaic blocks, 215
Multiplicity factor, 256, 257, 272

N

NaCl structure, 79, 81, 119, 165, 166, 270–273
 NaCl lattice, 270
NiAs structure, 79
Nonsymmetric scattering, 126
Number of broken bonds, 71, 91

O

Oblique P-lattice, 275
Octahedral site, 70
Off-Bragg angle diffraction, 134
Orthogonal, 6, 37, 43, 53, 60, 100, 210, 222, 277

Orthorhombicsystem, 35–39, 56, 60, 61, 81, 260, 277, 278
 lattices, 278
Oscillator model, 106, 107, 114
Out-of-plane orientation, 183–185, 191, 198, 201, 221–223

P

Penetration depth, 4, 12, 21, 26, 216
Periodic structures, 51, 108
 periodically arranged apertures, 108
 refractive-index variation, 108
 surface relief pattern, 108
Perovskite structure, 81, 82
Photo-absorption process, 24
Photographic film, 142, 226, 252, 255
Photons, 10, 12–14, 16, 23, 26, 154
Plane waves, 100–102, 112
Planes, 86
 (001)-planes, 56, 65, 73, 123, 125, 140, 148, 149, 159, 160, 171, 189, 257
 (001)-projection, 73, 78, 85, 90, 91, 271
 (010)-planes, 56, 257
 {101}, 197, 198
 (101), 90, 192, 193, 197, 198, 201
 {111}, 69, 72, 75, 79, 194, 195, 197, 198, 257
 (111)-oriented MgO, 183
 (111), 48, 62, 64, 65, 68–72, 78, 80, 89, 90, 133, 163, 169, 172, 177, 183, 190, 193, 194, 195, 196, 197, 201, 222, 257
Plank constant, 13
Plasma frequency, 11, 12
Plastic deformation, 75, 203
Point groups, 86, 88
Poisson's ratio, 204, 209
Pole, 86, 88, 193, 194
Pole of plane, 86
Poles of cubic crystal, 87
Polycrystalline flim, 184, 185, 191, 199, 200, 218, 250, 253
 film, 185, 199, 218

Index 287

Powder diffraction file, 251, 255, 261
  fingerprint, 251
  International Center for Diffraction Data, 251
Powder peak, 256, 257
Powder X-ray diffraction, 255
Powerful nondestructive technique, 182, 183
Primitive cell, 30, 40, 42, 76, 77, 89, 280
Principle of,
  Laue diffraction, 230
  powder diffraction, 253

R

Radio waves, 4
Reciprocal lattice, 28, 51, 52, 54–63, 66, 76, 77, 89, 130, 131, 139–143, 151, 171–175, 177, 187, 189, 194, 195, 198, 202, 215, 231, 232, 237, 238, 242, 245, 253
  Bragg law, 51
  vector analysis, 52
  X-ray diffraction, 51
Reciprocal lattice rods, 141, 142
Reciprocal lattice theory, 51
Reciprocal-space treatment of Laue diffraction, 230
Rectangular C-lattice, 275, 277
Rectangular P-lattice, 275, 277
Refraction, 38, 93–96, 102, 106, 107, 114
  phenomenon, 94
    law of energy, 94
    momentum conservation, 94
  refraction of light, 94
    glass prism, 94
    rainbows, 94
    white light into a rainbow-spectrum, 94
Refractive index, 50, 51, 94–96, 102, 104, 109, 113
Repeat unit, 29–31, 40, 44, 157
Resultant intensity, 99, 100
Rhombohedral cell, 30, 37, 40–42, 61
Rhombohedron, 37, 40, 42, 280

Rocking curve, 210, 214, 215
Rotational energy bands, 9, 10
  rotational resonances, 10
Rotational symmetryelements, 32, 35, 37, 40, 275, 279
Rotation-inversion, 32, 34–37
  axis, 34–36
  symmetry, 35

S

Scanning modes, 187
Scattering by an electron, 151
Scattering centers, 110–112, 127, 134
Scattering, structure factor, 157
Scattering, unit cell, 157
Scherrer equation, 137, 138
Scherrer formula, 138, 139
Second-order reflection, 123
Simple lattice, 30, 39, 43, 44, 72
Simple triclinic lattice, 275
Sinusoidal function, 6
Snell's law, 95, 102, 103, 107
Snell's law of refraction, 95, 107
Soft X-rays, 4
Solid solutions, 82–84, 259
Soller slits, 219
Southern hemisphere, 86
Space lattice, 29–31, 39, 40, 42, 43, 275, 279
Space lattices, 38, 277, 279
  based-centered, 277
  body-centered, 277
  face-centered, 277
  simple, 277
Space-vehicle communications, 9
Spectrum of, electromagnetic radiation, 9
Sphalerite structure, 80
Sphere of projection, 86, 87, 192, 193
Spherical wave, 97, 101, 105
Spinel structure, 82
Square P-lattice, 275
Stacking sequences for monoclinic P and C lattices, 277
Stacking unit cells, 31

Stereographic diagrams, 34
Stereographic projection, 86, 87, 227
Structure factor, 156, 162–173, 175, 177, 189, 195, 196, 201, 256, 259, 266, 270–272
Super-cooled liquid, 28
Superposition of waves, 160, 161
Surface energy, 71
Surface relief pattern, 108
Suspicion, 259
Symmetry, 28–45, 48–51, 61, 67, 73, 74, 83, 86–88, 155, 162, 172, 193–195, 197, 200, 202, 226–230, 236–244, 247, 256–258, 264, 275–280
Symmetry elements, 28, 32, 34, 35, 172, 238
   inversion, 32
   reflection, 32
   rotation-inversion, 32, 34–37
   rotation, 32
Systematic absence, 151, 171–173, 189, 191, 196, 266

T

Tensile stresses, 208, 209
Tetragonal, 35–39, 43–45, 49, 56, 60–62, 81, 86–90, 133, 144, 176, 193–196, 199, 221, 222, 226–228, 238, 239, 243–246, 255, 257, 260, 274, 278, 279
   crystal, 38, 193
Tetrahedral sites, 70, 71, 73, 78, 80, 82
Thick film, 182
Thin film, 5, 64, 104, 112, 113, 139–141, 182–184, 187–190, 193, 199–201, 204–212, 215, 216, 221–223, 255
Thomson equation, 153, 154
Three-beam interference, 113
Transmission electron microscope, 141
Transmission Laue method, 234, 235
Transmission method, 235, 236
Transmission spectrum, 113
Triclinic, 35, 36, 39, 276
Triequiangular layers, 42

Trigonalsystem, 35–42, 49, 61, 220, 239, 280
   crystal, 40
Tungsten, 15, 229
Two-dimensional (2D) lattices, 275, 276
   hexagonal P-lattice, 275, 276
   oblique P-lattice, 275, 276
   rectangular C-lattice, 275, 276
   rectangular P-lattice, 275, 276
   square P-lattice, 275, 276
Two-dimensional patterns, 33

U

Ultraviolet, 8
Uniaxial crystals, macroscopic properties of, 38
   anisotropic, 38
      electrical conduction, 38
      optical refraction, 38
      thermal expansion, 38
Unit cell, 29–31, 36–45, 47, 49, 54–66, 69–85, 88–91, 127, 148–151, 157–159, 162–173, 176, 177, 182, 220, 221, 236, 237, 245, 250–252, 255, 258–260, 264–266, 269–271, 275, 277, 280
Upright tetrahedron, 71

V

Vector analysis, 52
Visible light, 4, 8, 11, 14, 21, 26, 118, 121, 129
Von Laue's experiment, 226

W

Wave propagation, 6
Wave vectors, 101, 103, 112, 129
Wavefront, 97, 98
Wavelengths, 4, 8, 9, 11, 16, 17, 19, 24, 25, 96, 109, 119, 121, 145, 230–232, 235, 238, 242, 250
Wave-particle duality, 12
White radiation, 16, 17, 19, 226, 233, 236

Index 289

continuous radiation, 16
Wurtzite structure, 80, 91

X

X-ray crystallography, 118
X-ray diffraction, 5, 18, 19, 23, 24, 26,
51, 52, 83, 84, 102, 109, 118, 119,
121, 124, 126, 129, 134, 136, 137,
139, 141, 142, 145, 153, 154, 156,
162, 182, 183, 186, 203, 208, 215,
216, 219, 229, 233, 250, 253, 255,
256, 264
X-ray instrumentation, 261
X-ray tube, 15, 16, 19, 25, 152, 226,
229, 231, 236, 250, 255, 272
anode metal electrode, 15
cathode metal electrode, 15
X-rays, 3, 4, 8, 11, 12, 15, 16, 23–26,
96, 109, 111, 118–122, 124, 128,
131–135, 141, 144, 151, 153, 182,
209, 218, 226, 230, 231, 235, 238,
239, 247, 255, 258, 263

electromagneticradiations, 4
powerful for analyzing the internal
state of crystalline materials, 4
visuallyopaqueobjects, 4
computed tomography, 4
medical radiography, 4
security scanners, 4
wavelengths, 4
grating period, 4
hard X-rays, 4
soft X-rays, 4

Y

Young's modulus, 203, 204

Z

Zinc blende structure, 80, 166
sphalerite structure, 166
ZnS structure and its (001)-projection,
271
Zone axis, 59, 90, 142, 237, 238